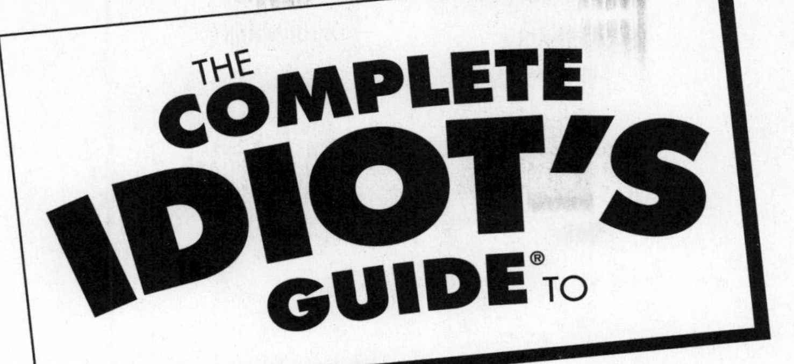

Algebra

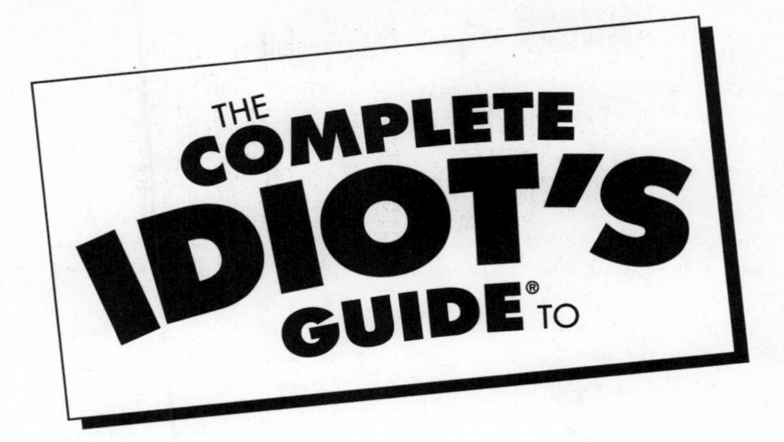

Algebra

by W. Michael Kelley

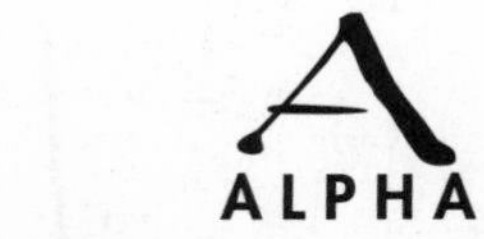

A member of Penguin Group (USA) Inc.

For my wife, Lisa, who makes my life worth living, and my son, Nicholas, who taught me that waking up in the morning with the people you love is just the best thing in the world.

ALPHA BOOKS

Published by the Penguin Group

Penguin Group (USA) Inc., 375 Hudson Street, New York, New York 10014, U.S.A.

Penguin Group (Canada), 10 Alcorn Avenue, Toronto, Ontario, Canada M4V 3B2 (a division of Pearson Penguin Canada Inc.)

Penguin Books Ltd, 80 Strand, London WC2R 0RL, England

Penguin Ireland, 25 St Stephen's Green, Dublin 2, Ireland (a division of Penguin Books Ltd)

Penguin Group (Australia), 250 Camberwell Road, Camberwell, Victoria 3124, Australia (a division of Pearson Australia Group Pty Ltd)

Penguin Books India Pvt Ltd, 11 Community Centre, Panchsheel Park, New Delhi—10 017, India

Penguin Group (NZ), cnr Airborne and Rosedale Roads, Albany, Auckland 1310, New Zealand (a division of Pearson New Zealand Ltd)

Penguin Books (South Africa) (Pty) Ltd, 24 Sturdee Avenue, Rosebank, Johannesburg 2196, South Africa

Penguin Books Ltd, Registered Offices: 80 Strand, London WC2R 0RL, England

International Standard Book Number: 1-59257-161-1
Library of Congress Catalog Card Number: 2004103222

08 07 06 8 7 6

Interpretation of the printing code: The rightmost number of the first series of numbers is the year of the book's printing; the rightmost number of the second series of numbers is the number of the book's printing. For example, a printing code of 04-1 shows that the first printing occurred in 2004.

Printed in the United States of America

Publisher: *Marie Butler-Knight*
Product Manager: *Phil Kitchel*
Senior Managing Editor: *Jennifer Chisholm*
Senior Acquisitions Editor: *Mike Sanders*
Development Editor: *Nancy D. Lewis*
Senior Production Editor: *Billy Fields*
Copy Editor: *Amy Borrelli*
Illustrator: *Richard King*
Cover/Book Designer: *Trina Wurst*
Indexer: *Brad Herriman*
Layout/Proofreading: *Becky Harmon, Donna Martin*

Contents at a Glance

Contents

Appendixes

Foreword

Just when you were getting used to dealing with numbers, along comes algebra! "Suddenly all these *x*'s and *y*'s start sprouting up all over"—as Mike Kelley so eloquently puts it—"like pimples on prom night." But have no fear, *The Complete Idiot's Guide to Algebra* is here!

As a naive junior in Mike Kelley's high school classroom, I daily looked forward to hearing his relevant and entertaining spin on math. I learned without even realizing it. His laid-back and humorous attitude—which is evident throughout the book—gently eases students into new and complex material.

Now that I am an English major at the University of Maryland, Baltimore County, I am no longer required to study math. So, why am I writing about this book? Because it works, that's why! Mr. Kelley provides information from the general to the specific, allowing you to see the forest for the trees and vice versa. From basic terms to complex equations, he is there each step of the way, guiding you with examples that are relevant to any high school or college student's life.

A few things that are going to be particularly helpful to the "mathematically challenged" are Mike's side notes. The "Kelley's Cautions" and "Critical Points" that frequent the sides of the pages contain valuable tips and hints to help you to easily assimilate new and involved subject matter. Then there are always "The Least You Need to Know" sections at the end of each chapter to assist you in deciding whether you are ready to move on or if you need to revisit some key ideas. And just so you don't get too bogged down in all of this mathematical mumbo jumbo, Mike constantly attempts to amuse you with his quirky jokes and anecdotes. It's like having your own personal Mr. Kelley in a box … or book, as it were.

But seriously, folks, this book is a practical and fun way to approach algebra. During his stint as a high school teacher, Mr. Kelley was in contact with many different levels of mathematical ability and many different attitudes about math. His experiences enable him to cater to them all. Even the most skilled mathematicians could learn something from Mike Kelley. Yet he still takes the time to explain the things that these algebraic geniuses take for granted. It's like he's talking *with* you, not at you or down to you. Mr. Kelley earnestly cares about your personal achievement in algebra and even in life, and that is the best thing about this book and about him as a person.

Good luck with algebra and in gaining the knowledge you need to move on to even more daring subjects like calculus, for which Mike Kelley also has a book (wink wink).

Enjoy!

Becky Reyno, former student

Introduction

Picture this scene in your mind. I am a high school student, chock-full of hormones and sugary snack cakes, thanks to puberty and the fact that I just spent the $3 my mom gave me for a healthy lunch on Twinkies and doughnuts in the cafeteria. I am young enough that I still like school, but old enough to understand that I'm not supposed to act like it, and my mind is active, alert, and tuned in. There are only two more classes to go and my day is over, and with that in mind, I head for algebra class.

In retrospect, I think the teacher must have had some sort of diabolical fun-sucking and joy-destroying laser ray gun hidden in the drop-down ceiling of that classroom, because just walking into algebra class put me in a bad mood. It's as hot as a varsity football player's armpit in that windowless, dank dungeon, and strangely enough, it always smells like a roomful of people just finished jogging in place in there. Vague yet acrid sweat and body odor attack my senses, and I slink down into my chair.

"I have to stay awake today," I tell myself. "I am on the brink of getting hopelessly lost, so if I drift off again, I won't understand anything, and we have a big test in a few days." However, no matter how I chide and cajole myself into paying attention, it is utterly impossible.

The teacher walks in and turns on a small oscillating fan in a vain effort to move the stinky air around and revive her class. Immediately she begins, in a soft, soothing voice, and the world in my peripheral vision begins to blur. Uh oh, soft monotonous vocal delivery, the droning white noise of a fan, the compelling malodorous warmth that only occupies rooms built out of brightly painted cinderblock … all elements that have thwarted my efforts to stay awake in class before.

I look around the room, and within 10 minutes, the majority of students are asleep. The rest are writing notes to boyfriends or girlfriends, and the school's star soccer player sits next to me, eyes wide and staring at his Trapper Keeper notebook, apparently having regressed into a vegetative state as soon as class began. I began to chant my daily mantra to myself, "I hate this class, I hate this class, I hate this class …" and I really meant it. To me, algebra is the most boring thing that was ever created, and it exists solely to destroy my happiness.

Can you relate to that story? Even though the individual details may not match your experience, did you possess a similar mantra? Some people have a hard time believing that a math major really hated math during his formative years. I guess the math after algebra got more interesting, or my attention span widened a little bit. However, that's not the normal course of events. Luckily, my extremely bad experience with math didn't prevent me from taking more classes, and eventually my opinion changed, but most people hit the brick wall of algebra and give up on math forever, in hopeless despair.

That was when I decided to go back and revisit the horribly boring and difficult mathematics classes I took, and write books that would not only explain things more clearly, but make a point of speaking in everyday language. Besides, I have always thought learning was much more fun when you could laugh along the way, but that's not necessarily the opinion of most math people. In fact, one of the mathematicians who reviewed my book *The Complete Idiot's Guide to Calculus* before it was released told me, "I don't think your jokes are appropriate. Math books shouldn't contain humor, because the math inside is already fun enough."

I believe that logic is insane. In this book, I've tried to present algebra in an interesting and relevant way, and attempted to make you smile a few times along the way. I didn't want to write a boring textbook, but at the same time, I didn't want to write an algebra joke book so ridiculously crammed with corny jokes that it insults your intelligence.

I also tried to include as much practice as humanly possible for you without making this book a million pages long. (Such books are hard to carry and tend to cost too much; besides, you wouldn't believe how expensive the shipping costs are if you buy them online!) Each section contains fully explained examples and practice problems to try on your own in little sidebars labeled "You've Got Problems." Additionally, Chapter 20 is jam-packed with practice problems based on the examples throughout the book, to help you identify your weaknesses if you've taken algebra before, or to test your overall knowledge once you've worked your way through the book. Remember, it doesn't hurt to go back to your algebra textbook and work out even more problems to hone your skills once you've exhausted the practice problems in this book, because repetition and practice transforms novices into experts.

Algebra is not something that can only be understood by a few select people. You can understand it and excel in your algebra class. Think of this book as a personal tutor, available to you 24 hours a day, seven days a week, always ready to explain the mysteries of math to you, even when the going gets rough.

How This Book Is Organized

This book is presented in seven sections:

In **Part 1, "A Final Farewell to Numbers,"** you'll firm up all of your basic arithmetic skills to make sure they are finely tuned and ready to face the challenges of algebra. You'll calculate greatest common factors and least common multiples, review exponential rules, tour the major algebraic properties, and explore the correct order of operations.

In **Part 2, "Equations and Inequalities,"** the preparation is over, and it's time for full-blown algebra. You'll solve equations, draw their graphs, create linear equations, and even investigate inequality statements with one and two variables.

In **Part 3, "Systems of Equations and Matrix Algebra,"** you'll find common solutions to multiple equations simultaneously. In addition, you'll be introduced to matrix algebra, a comparatively new branch of algebra that's really caught on since the dawn of the computer age.

Things get a little more intense in **Part 4, "Now You're Playing with (Exponential) Power!"** because the exponents are no longer content to stay small. You'll learn to cope with polynomials and radicals, and even how to solve equations that contain variables raised to the second, third, and fourth powers.

Part 5, "The Function Junction," introduces you to the mathematical function, a little machine that will take center stage as you advance in your mathematical career. You'll learn how to calculate a function's domain and range, find its inverse, and even graph it without having to resort to a monotonous and repetitive table of values.

Fractions are back in the spotlight in **Part 6, "Please, Be Rational!"** You'll learn how to do all the things you used to do with simple fractions (like add, subtract, multiply, and divide them) when the contents of the fractions get more complicated.

Finally, in **Part 7, "Wrapping Things Up,"** you'll face algebra's playground bully, the word problem. However, once you learn a few approaches for attacking word problems head on, you won't fear them anymore. You'll also get a chance to practice all of the skills covered throughout the book.

Things to Help You Out Along the Way

As a teacher, I constantly found myself going off on tangents—everything I mentioned reminded me of something else. These peripheral snippets are captured in this book as well. Here's a guide to the different sidebars you'll see peppering the pages that follow.

You've Got Problems

Math is not a spectator sport! Once we discuss a topic, I'll explain how to work out a certain type of problem, and then you have to try it on your own. These problems will be very similar to those that I walk you through in the chapters, but now it's your turn to shine. You'll find all the answers, explained step-by-step, in Appendix A.

Critical Point

These notes, tips, and thoughts will assist, teach, and entertain. They add a little something to the topic at hand, whether it be some sound advice, a bit of wisdom, or just something to lighten the mood a bit.

Talk the Talk

Algebra is chock-full of crazy- and nerdy-sounding **words** and **phrases.** In order to become King or Queen Math Nerd, you'll have to know what they mean!

Kelley's Cautions

Although I will warn you about common pitfalls and dangers throughout the book, the dangers in these boxes deserve special attention. Think of these as skulls and crossbones painted on little signs that stand along your path. Heeding these cautions can sometimes save you hours of frustration.

How'd You Do That?

All too often, algebraic formulas appear like magic, or you just do something because your teacher told you to. If you've ever wondered "Why does that work?", "Where did that come from?", or "How did that happen?", this is where you'll find the answer.

Acknowledgments

If I have learned anything in the short time I've spent as an author, it's that authors are insecure people, needing constant attention and support from friends, family members, and folks from the publishing house, and I lucked out on all counts. Special thanks are extended to my greatest supporter, Lisa, who never growled when I trudged into my basement and dove into my work, day in and day out (and still didn't mind that I watched football all weekend long—honestly, she must be the world's greatest wife). Also, thanks to my extended family and friends, especially Dave, Chris, Matt, and Rob, who never acted like they were tired of hearing every boring detail about the book as I was writing.

Thanks go to my agent, Jessica Faust at Bookends, LLC, who pushed and pushed to get me two great book-writing opportunities, and Nancy Lewis, my development editor, who is eager and willing to put out the little fires I always end up setting every day. Also, I have to thank Mike Sanders at Pearson/Penguin, who must have tons of experience with neurotic writers, because he's always so nice to me.

Sue Strickland, my mentor and one-time college instructor, once again agreed to technically review this book, and I am indebted to her for her direction and expertise. Her love of her students is contagious, and it couldn't help but rub off on me.

Here and there throughout this book, you'll find in-chapter illustrations by Chris Sarampote, a longtime friend and a magnificent artist. Thanks, Chris, for your amazing drawings, and your patience when I'd call in the middle of the night and say "I think the arrow in the football picture might be too curvy." Visit him at www.sarampoteweb.com, if you dare.

Finally, I need to thank Daniel Brown, my high school English teacher, who one day pulled me aside and said "One day, you will write math books for people such as I, who approach math with great fear and trepidation." His encouragement, professionalism, and knowledge are most of the reason that his prophecy has come true.

Special Thanks to the Technical Reviewer

The Complete Idiot's Guide to Algebra was reviewed by an expert who double-checked the accuracy of what you'll learn here, to help us ensure that this book gives you everything you need to know about algebra. Special thanks are extended to Susan Strickland, who also provided the same service for *The Complete Idiot's Guide to Calculus.*

Susan Strickland received a Bachelor's degree in mathematics from St. Mary's College of Maryland in 1979, a Master's degree in mathematics from Lehigh University in 1982, and took graduate courses in mathematics and mathematics education at The American University in Washington, D.C., from 1989 through 1991. She was an assistant professor of mathematics and supervised student teachers in secondary mathematics at St. Mary's College of Maryland from 1983 through 2001. It was during that time that she had the pleasure of teaching Michael Kelley and supervising his student teacher experience. Since 2001, she has been a professor of mathematics at the College of Southern Maryland and is now involved with teaching math to future elementary school teachers. Her interests include teaching mathematics to "math phobics," training new math teachers, and solving math games and puzzles (she can really solve the Rubik's Cube).

Trademarks

Part 1

A Final Farewell to Numbers

When most people think of math, they think "numbers." To them, math is just a way to figure out how much they should tip their waitress. However, math is so much more than just a substitute for a laminated card in your wallet that tells you what 15 percent of the price of dinner should be. In this part, I am going to make sure you're up to speed with numbers, and have mastered all of the basic skills you'll need later on.

Chapter 1

Getting Cozy with Numbers

In This Chapter

- Categorizing types of numbers
- Coping with oodles of signs
- Brushing up on prealgebra skills
- Exploring common mathematical assumptions

Most people new to algebra view it as a disgusting, creeping disease whose sole purpose is to ruin everything they've ever known about math. They understand multiplication, and can even divide numbers containing decimals (as long as they can check their answers with a calculator or a nerdy friend), but algebra is an entirely different beast—it contains *letters!* Just when you feel like you've got a handle on math, suddenly all these *x*'s and *y*'s start sprouting up all over, like pimples on prom night.

Before I can even begin talking about those letters (they're actually called *variables*), you've got to know a few things about those plain old numbers you've been dealing with all these years. Some of the things I'll discuss in this chapter will sound familiar, but most likely some of it will also be new. In essence, this chapter will be a grab bag of prealgebra skills I need to review with you; it's one last chance to get to know your old number friends better, before we unceremoniously dump letters into the mix.

Classifying Number Sets

Most things can be classified in a bunch of different ways. For example, if you had a cousin named Scott, he might fall under the following categories: people in your family, your cousins, people with dark hair, and (arguably) people who could stand to brush their teeth a little more often. It would be unfair to consider only Scott's hygiene (lucky for him); that's only one classification. A broader picture is painted if you consider all of the groups he belongs to:

- People in your family
- Your cousins
- People with dark hair
- Hygienically challenged people

The same goes for numbers. Numbers fall into all kinds of categories, and just because they belong to one group, it does not preclude them from belonging to others as well.

Familiar Classifications

You've been at this number classification thing for some time now. In fact, the following number groups will probably ring a bell:

Talk the Talk

If a number is **evenly divisible** by 2, that means if you divide that number by 2, there will be no remainder.

Critical Point

Technically, 0 is divisible by 2, so it is considered even. However, 0 is not positive, nor is it negative—it's just sort of hanging out there in mathematical purgatory, and can be classified as both *nonpositive* and *nonnegative*.

- *Even numbers:* Any number that's *evenly divisible* by 2 is an even number, like 4, 12, and –10.
- *Odd numbers:* Any number that is not evenly divisible by 2 (in other words, when you divide by 2, you get a remainder) is an odd number, like 3, 9, and –25.
- *Positive numbers:* All numbers greater than 0 are considered positive.
- *Negative numbers:* All numbers less than 0 are considered negative.
- *Prime numbers:* The only two numbers that divide evenly into a prime number are the number itself and 1 (and that's no great feat, since 1 divides evenly into every number). Some examples of prime numbers are 5, 13, and 19. By the way, 1 is not considered a prime number, due to the technicality

that it's only divisible by one thing, while all the other prime numbers are divisible by two things. Therefore, 2 is the smallest positive prime number.

- *Composite numbers:* If a number is divisible by things other than itself and 1, then it is called a composite number, and those things that divide evenly into the number (leaving behind no remainder) are called its *factors.* Some examples of composite numbers are 4, 12, and 30.

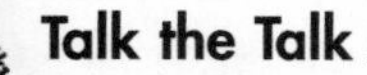

Talk the Talk

A **factor** is a number that divides evenly into a number and leaves behind no remainder. For example, the factors for the number 30 are 1, 2, 3, 5, 6, 10, 15, and 30.

I don't mean to insult your intelligence by reviewing these simple categories. Instead, I mean to instill a little confidence before I start discussing the slightly more complicated classifications.

Intensely Mathematical Classifications

Math historians (if you thought regular math people were boring, you should get a load of these guys) generally agree that the earliest humans on the planet had a very simple number system that went like this: one, two, a lot. There was no need for more numbers. Lucky you—that's not true any more. Here are the less familiar number classifications you'll need to understand:

- *Natural numbers:* The numbers 1, 2, 3, 4, 5, and so forth are called the natural (or *counting*) numbers. They're the numbers you were first taught as a child when counting.

- *Whole numbers:* Throw in the number 0 with the natural numbers and you get the whole numbers. That's the only difference—0 is a whole number but not a natural number. (That's easy to remember, since a 0 looks like a drawing of a hole.)

- *Integers:* Any number that has no explicit decimal or fractional component is an integer. Therefore, –4, 17, and 0 are integers, but 1.25 and $\frac{2}{5}$ are not.

- *Rational numbers:* If a number can be expressed as a decimal that either repeats infinitely or simply ends (called a *terminating decimal*), then the number is rational. Basically, either of those conditions guarantees one thing: That number is actually equivalent to a fraction, so all fractions are automatically rational.

  (You can remember this using the mnemonic device "Rational means fractional." The words sound roughly the same.) Using this definition, it's easy to see that the numbers $\frac{1}{3}$, 7.95, and .8383838383… are all rational.

- *Irrational numbers:* If a number cannot be expressed as a fraction, or its decimal representation goes on and on infinitely but not according to some obvious repeating pattern of digits, then the number is irrational. Although many radicals (square roots, cube roots, and the like) are irrational, the most famous irrational number is $\pi = 3.141592653589793\ldots$. No matter how many thousands (or millions) of decimal places you examine, there is no pattern to the numbers. In case you're curious, there are far more irrational numbers that exist than rational numbers, even though the rationals include every conceivable fraction!
- *Real numbers:* If you clump all of the rational and irrational numbers together, you get the set of real numbers. Basically, any number that can be expressed as a single decimal (whether it be repeating, terminating, attractive, or awkward-looking but with a nice personality) is considered a real number.

Critical Point

Since every integer is divisible by 1, that means each can be written as a fraction ($7 = \frac{7}{1}$). Therefore, every integer is also a rational number.

Don't be intimidated by all the different classifications. Just mark this page, and check back when you need a refresher.

Example 1: Identify the categories that the number 8 belongs to.

Solution: Since there's no negative sign preceding it, 8 is a positive number. Furthermore, it has factors of 1, 2, 4, and 8 (since all those numbers divide evenly into 8), indicating that 8 is both even and composite. Additionally, 8 is a natural number, a whole number, an integer, a rational number $\left(\frac{8}{1}\right)$, and a real number (8.0).

You've Got Problems

Problem 1: Identify all the categories that the number $\frac{3}{7}$ belongs to.

Persnickety Signs

Before algebra came along, you were only expected to perform operations (such as addition or multiplication) on positive integers, but now you'll be expected to perform those same operations on negative numbers as well. So, before you get knee-deep in fancy algebra techniques, it's important that you understand how to work with positives and negatives at the same time. The procedures you'll use for addition and subtraction are completely different than the ones for multiplication and division, so I'll discuss those separately.

Addition and Subtraction

On the first day of one of my statistics courses in college, the professor started by asking us, "What is 5 – 9?" The answer he expected, of course, was –4. However, the first student to raise his hand answered unexpectedly. "That's impossible," he said, "You can't take 9 apples away from 5 apples—you're out of apples!" Keep in mind that this was a college senior, and you can begin to understand the despair felt by the professor. It's hard to learn high-level statistics when a student doesn't understand basic algebra.

Here's some advice: Don't think in terms of apples, as tasty as they may be. Instead, think in terms of earning and losing money—that's something everyone can relate to, and it makes adding and subtracting positive and negative numbers a snap. If, at the end of the problem, you have money left over, your answer is positive. If you're short on cash and still owe, your answer is negative.

> CAUTION
>
> **Kelley's Cautions**
>
> Most textbooks write negative numbers like this: –3. However, some write the negative sign way up high like this: ¯3. Both notational methods mean the exact same thing, although I won't use that weird, sky-high negative sign.

Example 2: Simplify $5 - (-3) - (+2) + (-7)$.

Solution: This is the perfect example of an absolutely evil addition and subtraction problem, but if you follow two simple steps, it becomes quite simple.

1. **Eliminate double signs (signs that are not separated by numbers).** If two consecutive signs are the same, replace them with a single positive sign. If the two signs are different, replace them with a single negative sign.

Ignore the parentheses for a minute and work left to right. You've got two negatives right next to each other (between the 5 and 3). Since those consecutive signs are the same, replace them with a positive sign. The other two pairs of consecutive signs (between the 3 and 2 and then between the 2 and 7) are different, so they get replaced by a negative sign:

$$5 + 3 - 2 - 7$$

Once double signs are eliminated, you can move on to the next step.

2. **Consider all positive numbers as money you earn and all negative numbers as money you lose to calculate the final answer.** Remember, if there is no sign immediately preceding a number, that number is assumed to be positive. (Like the 5 in this example.)

You can read the problem 5 + 3 – 2 – 7 as "I earned five dollars, then three more, but then lost two dollars and then lost seven more." You end up with a total net loss of one dollar, so your answer is –1.

Notice that I don't describe different techniques for addition and subtraction; this is because subtraction is actually just addition in disguise—it's basically just adding negative numbers.

You've Got Problems

Problem 2: Simplify 6 + (+2) – (+5) – (–4).

Multiplication and Division

When multiplying and dividing positive and negative numbers, all you have to do is follow the same "double signs" rule of thumb that you used in addition and subtraction, with a slight twist. If two numbers you're multiplying or dividing have the same sign, then the result will be positive, but if they have different signs, the result will be negative. That's all there is to it.

Example 3: Simplify the following:

(a) 5 × (–2)

Solution: Since the 5 and the 2 have different signs, the result will be negative. Just multiply 5 times 2 and slap a negative sign on your answer: –10.

(b) –18 ÷ (–6)

Solution: In this problem, the signs are the same, so the answer will be positive: 3.

You've Got Problems

Problem 3: Simplify the following:

(a) –5 × (–8)

(b) –20 ÷ 4

Opposites and Absolute Values

There are two things you can do to a number that may or may not change its sign: Calculate its opposite and calculate its absolute value. Even though these two things have similar purposes (and are often confused with one another), they work in entirely different ways.

The *opposite* of a number is indicated by a lone negative sign out in front of it. For example, the opposite of –3 would be written like this: –(–3). The value of a number's opposite is simply the number multiplied by –1. Therefore, the only difference between a number and its opposite is its sign.

$$-\left(-\frac{1}{2}\right)=\frac{1}{2} \qquad -(4)=-4$$

On the other hand, the *absolute value* of a number doesn't always have a different sign than the original number. Absolute values are indicated by thin vertical lines surrounding a number like this: $|-9|$. (You read that as "the absolute value of –9.")

What's the purpose of an absolute value? It always returns the positive version of whatever's inside it. Absolute value bars are sort of like "instant negative sign removers," and are so effective they should have their own infomercial on TV. ("Does your laundry have stubborn negative signs in it that just *refuse* to come out?") Therefore, $|-3|$ is equal to 3.

Notice that the absolute value of a positive number is also positive! For example, $|21|=21$. Since absolute values only take away negative signs, if the original number isn't negative, they don't have any effect on it at all.

Talk the Talk

The **opposite** of a number has the same value but the opposite sign of that number. The **absolute value** of a number has the same value but will always be positive.

You've Got Problems

Problem 4: Determine the values of –(8) and $|8|$.

Come Together with Grouping Symbols

The absolute value symbols you were just introduced to are just one example of algebraic *grouping symbols*. Other grouping symbols include parentheses ( ), brackets [ ], and braces {}. These symbols surround all or portions of a math problem, and whatever appears inside the symbols is considered grouped together.

Critical Point

Technically, a fraction bar is also a grouping symbol, because it separates a fraction into two parts, the numerator and denominator. Therefore, you should simplify the two parts separately at the beginning of the problem.

Grouping symbols are important because they often help clue you in on what to do first when simplifying a problem. Actually, there is a very specific order in which you are supposed to simplify things, called the *order of operations*, which I'll discuss in greater detail in Chapter 3. Until then, just remember that anything appearing within any type of grouping symbols should be done first.

Example 4: Simplify the following.

(a) $15 \div \{7 - 2\}$

Solution: Because 7 – 2 appears in braces, you should combine those numbers together before dividing:

$$15 \div 5 = 3$$

(b) $|5 - 3 + (-8)|$

Solution: Because absolute value bars are present, you may be tempted to strip away all the negative signs. However, since they are grouping symbols, you must first simplify inside them. Eliminate double signs and combine the numbers as you did earlier.

$$|5 - 3 - 8|$$

$$|2 - 8|$$

$$|-6|$$

Now that the content of the absolute values has been completely simplified, you can take the absolute value of –6 and get 6 for your final answer.

(c) $10 - [6 \times (2 + 1)]$

Solution: No grouping symbol has precedence over another. For example, you don't always do brackets before braces. However, if more than one grouping symbol appears in a problem, do the innermost set first, and work your way out.

In this problem, the parentheses are contained within another grouping symbol—the brackets—making the parentheses the innermost symbols. So, you should simplify 2 + 1 first.

$$10 - [6 \times 3]$$

Only one set of grouping symbols remains, the brackets. Go ahead and simplify their contents next.

$$10 - 18$$

All that's left between you and the joy of a final answer is a simple subtraction problem whose answer is –8.

You've Got Problems

Problem 5: Simplify the following:

(a) $5 \times [4 - 2]$

(b) $|2 - (16 \div 4)|$

Important Assumptions

Just because something has a very complicated name attached to it doesn't mean that the concept is necessarily very difficult to understand. Have you ever heard of hippopotomonstrosesquippedaliophobia?

From the root word "phobia," it's obviously a fear of some kind, and based on the length and complexity of the name, you might think it's some kind of powerfully debilitating fear with an intricate neurological or psychosocial cause. Maybe it's the kind of fear that's triggered by some sort of traumatic event, like discovering that your favorite television show has been preempted again by a presidential address. (That's my greatest fear, anyway.)

Actually, hippopotomonstrosesquippedaliophobia means "the fear of long words." In my experience, whether or not they begin the class with this fear, most algebra students develop it at some point during the course. You must fight it! Although the concepts I am about to introduce have rather strange and complicated names, they represent very simple ideas. Math people, like most professionals, just give complicated names to things they think are the most important.

Talk the Talk

An algebraic property (or axiom) is a mathematical fact that is so obvious, it is accepted without proof.

In this case, the important concepts are algebraic properties (or axioms), assumptions about the ways numbers work that cannot really be verified through technical mathematical proofs, but are so obviously true that math folks (who don't usually do such rash things) assume them to be true even with no hard evidence. Of course, you can show them to be true for any examples you may concoct (as I will when I discuss them), but you cannot prove them generically for any numbers in the world.

> **Kelley's Cautions**
>
> The four properties listed in this section (associative, commutative, identity, and inverse properties) are not the only mathematical axioms; in fact, two more are introduced in Chapter 3.

Your goal, when reading about these properties, is to be able to match the concept with the name, because you'll see the properties used later on in the book.

Associative Property

It's a natural tendency in people to split into social groups, so that they can spend more time with the people whose interests match their own. As a high school teacher, I had kids from all the cliques: the drama kids, the band kids, the jocks, the jerks; everyone was represented somewhere. However, no matter how they associated amongst themselves, as a group, the student pool stayed the same. The same is true with numbers.

No matter how numbers choose to associate within grouping symbols, their value does not change (at least with addition and multiplication, that is). Consider the addition problem

$$(3 + 5) + 9$$

The 3 and the 5 have huddled up together, leaving the poor 9 out in the cold, wondering if it's his aftershave to blame for his role as social pariah. If you simplify this addition problem, you should start inside the parentheses, since grouping symbols always come first.

$$8 + 9 = 17$$

If I leave the numbers in the exact same order but, instead, group the 5 and 9 together, the result will be the same.

$$3 + (5 + 9)$$

$$3 + 14 = 17$$

This is called the *associative property of addition;* in essence, it means that given a string of numbers to add together, it doesn't matter which you add first—the result will be the same. As I mentioned a moment ago, you also have an *associative property for multiplication*. Watch how, once again, differently placed grouping symbols do not affect the simplified outcome:

$$(2 \times 6) \times 4 = 2 \times (6 \times 4)$$

$$12 \times 4 = 2 \times 24$$

$$48 = 48$$

CAUTION

Kelley's Cautions

The operations of subtraction and division are *not* associative; different grouping symbol placements end in completely different results. Here's just one example proving that division is not associative:

$$(40 \div 10) \div 2 \neq 40 \div (10 \div 2)$$

$$4 \div 2 \neq 40 \div 5$$

$$2 \neq 8$$

Commutative Property

I have a hefty commute to work—it ranges between 75 and 120 minutes one way. One thing I notice all the time are these crazy drivers who whip in and out of lanes of traffic, just to move one or two car lengths further up the road. Even though they may get 10 or 20 feet ahead of you, a few minutes later, you usually end up passing them anyway. For all their dangerous stunt driving, they don't actually gain any ground. The moral of the story: No matter what the order of the commuters, generally, everyone gets to work at the same time. Numbers already know this to be true.

When you are adding or multiplying (once again, this property is not true for subtraction or division), the order of the numbers does not matter. Check out the multiplication problem

$$3 \times 2 \times 7$$

If you multiply left to right, $3 \times 2 = 6$, and then $6 \times 7 = 42$. Did you know that you'll still get 42 even if you scramble the order of the numbers? It's called the *commutative property of multiplication*. Need to see it in action? Here you go. (Don't forget to multiply left to right again.)

$$7 \times 3 \times 2 = 21 \times 2 = 42$$

Remember, there's also a *commutative property of addition:*

$$5 + 19 + 4 = 19 + 4 + 5$$

$$24 + 4 = 23 + 5$$

$$28 = 28$$

Kelley's Cautions

Here's one example that demonstrates why there's no commutative property for subtraction:

$$12 - 4 - 5 \neq 4 - 12 - 5$$

$$8 - 5 \neq -8 - 5$$

$$3 \neq -13$$

Identity Properties

Both addition and multiplication (poor subtraction and division—nothing works for them) have numbers called *identity elements*, whose job is (believe it or not) to leave numbers alone. That's right—their entire job is to make sure the number you start with doesn't change its identity by the time the problem's over.

The identity element for addition (called the *additive identity*) is 0, because if you add 0 to any number, you get what you started with.

$$3+0=3 \qquad \frac{1}{2}+0=\frac{1}{2} \qquad -37+0=-37$$

Pretty simple, eh? Can you guess what the multiplicative identity is? What is the only thing that, if multiplied by any number, will return the original number? The answer is 1—anything times 1 equals itself.

$$9 \times 1 = 9 \qquad 4 \times 1 = 4 \qquad -10 \times 1 = -10$$

These identity elements are used in the inverse properties as well.

Inverse Properties

The purposes of the inverse properties are to "cancel out" a number, and to get a final result that is equal to the identity element of the operation in question. That sounds complicated, but here's what it boils down to:

- **Additive Inverse Property:** Every number has an opposite (I discussed this a few sections ago) so that when you add a number to its opposite, the result is the additive identity element (0):

$$2 + (-2) = 0 \qquad -7 + 7 = 0$$

- **Multiplicative Inverse Property:** Every number has a *reciprocal* (defined as the fraction 1 over that number) so that when you multiply a number by its reciprocal, you get the multiplicative identity element (1):

$$5 \times \left(\frac{1}{5}\right) = 1 \qquad -6 \times \left(-\frac{1}{6}\right) = 1$$

That final property might be a bit troubling, because it requires that you know a thing or two about fractions. Don't worry if you get hung up on fractions, though. Chapter 2 deals exclusively with those nasty little fractions, and it'll help you get up to speed in case they cause you cold sweats and night terrors (like they do most people).

You've Got Problems

Problem 6: Name the mathematical properties that guarantee the following statements are true.

(a) $11 + 6 = 6 + 11$

(b) $-9 + 9 = 0$

(c) $(1 \times 5) \times 7 = 1 \times (5 \times 7)$

The Least You Should Know

- Numbers can be classified in many different ways, varying from their divisibility to whether or not you can write them as a fraction.
- Different techniques apply when adding and subtracting positive and negative numbers than if you were to multiply or divide them.
- You should always calculate the value of numbers located within grouping symbols first.
- Absolute value signs spit out the positive version of their contents.
- Mathematical properties are important (although unprovable) facts that describe intuitive mathematical truths.

Chapter 2

Making Friends with Fractions

In This Chapter

- Understanding what fractions are
- Writing fractions in different ways
- Simplifying fractions
- Adding, subtracting, multiplying, and dividing fractions

Few words have the innate power to terrify people like the word "fraction." It's quite a jump to go from talking about a regular number to talking about a weird number that's made up of two other numbers sewn together! Modern-day math teachers spend a lot of time introducing this concept to young students using toy blocks and educational manipulatives to physically model fractions, but some still stick to the old-fashioned method of teaching (like most of my teachers), which is to simply introduce the topic with no explanation, and then make you feel stupid if you have questions or don't understand.

In this chapter, I'll help you review your fraction skills, and I promise not to tease you if you have to reread portions of it a few times before you catch on. Throughout the book (and especially in Chapters 17 and 18), you'll be dealing with more complicated fractions that contain variables, so you should refine your basic fraction skills while there are still just numbers inside them.

What Is a Fraction?

There are three ways to think of fractions, all equally accurate, and each one gives you a different insight into what makes a fraction tick. In essence, a fraction is …

- **A division problem frozen in time.** A fraction is just a division problem written vertically with a fraction bar instead of horizontally with a division symbol; for example, you can rewrite $5 \div 7$ as $\frac{5}{7}$. Although they look different, those two things mean the exact same thing.

  Why, then, would you use fractions? Well, it's no big surprise that the answer to $5 \div 7$ isn't a simple number, like 2. Instead, it's a pretty ugly decimal value. To save yourself the frustration of writing out a ton of decimal places and the mental anguish of looking at such an ugly monstrosity, leave the division problem frozen in time in fraction form.

- **Some portion of a whole number or set.** As long as the top number in a fraction is smaller than its bottom number, the fraction has a (probably hideous-looking) decimal value less than one. "One what?" you may ask. It depends. For example, if you have seven eggs left out of the dozen you bought on Sunday at the supermarket, you could accurately say that you have $\frac{7}{12}$ (read "seven twelfths") of a dozen left. Likewise, since 3 teaspoons make up a tablespoon, if a recipe calls for 2 teaspoons, that amount is equal to $\frac{2}{3}$ (read "two thirds") of a tablespoon.

  When considering a fraction as a portion of a whole set, the top number represents how many items are present, and the bottom number represents how many items make one complete set.

- **A failed marketing attempt.** In the late 1700s, the popularity of mathematics in society began to wane, so in a desperate attempt to increase the popularity of numbers, scientists "supersized" them, creating fractions that included two numbers for the price of one. It failed miserably and mathematicians were forever shunned from polite society and forced to wear glasses held together by masking tape. By the way, this last one may not be true—I think I may have dreamed it.

Talk the Talk

The top part of a fraction is its **numerator** and the bottom part is the **denominator**.

By the way, the fancy mathematical name for the top part of the fraction is the *numerator*, and the bottom part is called the *denominator*. These terms are easily confused, so I have concocted a naughty way to help you remember which is which:

$$\frac{\textbf{Nu}\textit{merator}}{\textbf{De}\textit{nominator}}$$

As long as you read the first two letters of each word, from top to bottom, all of the mysteries of fractions will be laid "bare," so to speak.

Ways to Write Fractions

Remember, you can find the actual decimal value of a fraction if you divide the numerator by the denominator, effectively thawing out the frozen division problem. A calculator yields the answer fastest, but if you're one of those people who insists on doing things the old-fashioned way, long division works as well.

For example, the decimal value of $\frac{7}{12}$ is equal to $7 \div 12$, which is .5833333.... Note that the digit 3 will repeat infinitely. Any digit or digits in a decimal that behave like this can be written with a bar over them like so: $.58\overline{3}$. The bar just means "anything under here repeats itself over and over again."

Some fractions, called *improper fractions*, have numerators that are larger than their denominators, like $\frac{14}{5}$. Think of this fraction like a collection of elements (as I described in the previous section): You have 14 items, and it takes only 5 items to make one whole set. Therefore, you have enough for 2 sets (which would require 10) but not enough for 3 full sets (which would require 15). Therefore, the decimal value of $\frac{14}{5}$ is somewhere between 2 and 3 (but it's closer to 3 than 2). By the way, fractions whose numerators are smaller than their denominators are called *proper fractions*.

All improper fractions can be written as *mixed numbers*, which have both an integer and fraction part and make the actual value of the fraction easier to visualize. The improper fraction $\frac{14}{5}$ corresponds to the mixed number $2\frac{4}{5}$. Here's how I got that:

1. Divide the numerator by the denominator. (In this case, $14 \div 5$ divides in 2 times with 4 left over. Mathematically, we call 2 the *quotient* and 4 the *remainder*.)
2. The quotient will be the integer portion of the mixed number (the big number out front). The fraction part of the mixed number will be the reminder divided by the improper fraction's original denominator.

Talk the Talk

If the numerator is greater than the denominator, then you have an **improper fraction,** which can be left as is or transformed into a **mixed number,** with both integer and fraction parts. If the numerator is less than the denominator, it's a **proper fraction.**

Critical Point

To convert a mixed number into an improper fraction, all you have to do is add the integer part to the fraction part. In other words, $2+\frac{4}{5}=\frac{14}{5}$. If you don't already know how, I explain how to add fractions later in this chapter (see the section "Adding and Subtracting Fractions").

Most teachers would rather you leave your answers as an improper fraction, rather than express it as a mixed number, even though the implication of the term "improper fraction" might suggest that it is, in some way, wrong or a breach of manners to do so.

Simplifying Fractions

Did you know that fractions don't have to look the same to be equal? I wish this were true of humans as well, because then I might sometimes be confused with George Clooney. Alas, it is not true, and I am doomed to a life of comparison with a sea of other balding and soft-around-the-middle guys like me.

Talk the Talk

Once there are no factors common to both the numerator and denominator left in the fraction, the fraction is said to be in **simplified form.** The process of eliminating the common factors is called **simplifying** or **reducing** the fraction.

Talk the Talk

The largest factor two numbers have in common is called (quite predictably) the **greatest common factor,** and is abbreviated GCF.

Because equivalent (or equal) fractions can take on a whole host of forms, most instructors require you to put a fraction in *simplified form*, which means that its numerator and denominator don't have any factors in common. Every fraction has one unique simplified form, which you can reach by dividing out those common factors.

Example 1: Simplify the fraction $\frac{24}{36}$.

Solution: Do 24 and 36 have any factors in common? Sure—they're both even for starters, so they have a common factor of 2. Divide both parts of the fraction by 2 in an attempt to simplify it, and you get $\frac{12}{18}$. However, you're not done yet! This fraction can be simplified further, since both the numerator and denominator can be evenly divided by 6; divide both by that common factor to get $\frac{2}{3}$. Since 2 and 3 share no common factors other than 1, you're finished.

By the way, even though it took me two steps to simplify this fraction, you could have done it in one step, if you realized that 12 was the *greatest common factor* of 24 and 36. If you divide both numbers by the greatest common factor, you simplify the fraction in one step; in this case, you'll immediately get the answer of $\frac{2}{3}$.

If you're not convinced that the fractions $\frac{2}{3}$ and $\frac{24}{36}$ are just two different representations of the same value, convert them into decimals. It turns out they are both exactly equal to $.\bar{6}$, conclusive evidence that they are equivalent!

You've Got Problems

Problem 1: Simplify the fractions and identify the greatest common factor of the numerator and denominator.

(a) $\frac{7}{21}$

(b) $\frac{24}{40}$

Locating the Least Common Denominator

Sometimes, it's not useful to completely simplify fractions. In some cases (which I'll discuss in greater detail later in the chapter), you'll prefer fractions to look similar in appearance rather than be fully simplified. Specifically, you'll want to rewrite fractions so that they have the same denominator.

That's not as hard as it seems. Remember, fractions $\left(\text{like } \frac{2}{3} \text{ and } \frac{24}{36}\right)$ can look dramatically different but actually have the exact same value. The tricky part of rewriting fractions to have common (equal) denominators is figuring out exactly what that common denominator should be. However, if you follow these steps, identifying a common denominator should pose no challenge at all:

1. Examine the denominators in all of the fractions. Choose the largest of the group. For grins, I will call this large denominator Bubba.
2. Do all of the other, smaller denominators divide evenly into Bubba? If so, then Bubba is your least common denominator. If not, proceed to step 3.
3. Multiply Bubba by 2. Do all of the other denominators *now* divide evenly into Bubba? If not, multiply Bubba by 3 and see if that works. If not, continue multiplying Bubba by bigger and bigger numbers until all the denominators divide evenly into that big boy.

Not only does this procedure generate a common denominator, it actually generates the smallest one, called the least common denominator (abbreviated LCD).

CAUTION

Kelley's Cautions

Some students cop out when calculating a common denominator; rather than follow my simple three-step procedure, they multiply all of the denominators together. For example, to find a common denominator for the fractions $\frac{7}{200}$, $\frac{13}{100}$, and $-\frac{8}{50}$, they would calculate $200 \times 100 \times 50 = 1{,}000{,}000$, quite a monstrous common denominator!

Although that is a valid common denominator, it is certainly not the smallest one. According to my procedure, 200 is the largest denominator (hence it is named Bubba), and since the other denominators (100 and 50) both divide into Bubba evenly, then Bubba is the least common denominator. I don't know about you, but I'd much rather deal with a denominator of 200 than one that's 5,000 times as large!

Even though both processes will sometimes produce the exact same answer, my process always returns the best result, even if it's not quite as fast as the other.

Once you've calculated the least common denominator, you're halfway done. You still have to rewrite the fractions so that they contain their brand new, shiny common denominators. Here's how to do it.

1. Divide the least common denominator by the denominator of each fraction (you know for sure that it will divide evenly).
2. Multiply both the numerator and denominator of each fraction by the result.

I bet this whole process is a little intimidating. Indeed, it will take a little practice until the process becomes second nature to you, but (as you'll see in the next few examples) there's no one step that's overly complicated.

Example 2: Rewrite the fractions so that they contain the least common denominator.

(a) $\frac{1}{3}, \frac{7}{15}$

Solution: First, look for the biggest denominator (Bubba): 15. Since the other denominator (3) divides into 15 evenly, then 15 is the least common denominator, which is handy, because it means the second fraction won't need to be rewritten. However, the first fraction must be altered so that it also has a denominator of 15.

To rewrite the first fraction, begin by dividing the LCD by the denominator ($15 \div 3$). Now multiply the fraction's numerator and denominator by the result (5):

$$\frac{1 \times 5}{3 \times 5} = \frac{5}{15}$$

Notice that the new fraction is not in simplified form (if you were to simplify it, you'd get back what you started with, $\frac{1}{3}$).

Your final fractions, now written with common denominators, are $\frac{5}{15}$ and $\frac{7}{15}$.

How'd You Do That?

When you multiply the numerator and denominator of a fraction by the same number, it's actually the same thing as multiplying the fraction by a value of 1, which won't actually affect the fraction's value, since 1 is the multiplicative identity. That way the original $\left(\frac{1}{3}\right)$ and final fractions $\left(\frac{5}{15}\right)$ have the same value.

(b) $\frac{1}{2}, \frac{1}{3},$ and $\frac{3}{4}$

Solution: Bubba equals 4 in this set of fractions. Even though 2 divides into Bubba evenly, 3 does not. So, multiply Bubba times 2 (4×2) to get 8. Stubbornly, 3 still will not divide into 8 evenly. Oh well, onwards and upwards. Now, I multiply Bubba by 3 (4×3) to get 12. Finally, both of the other denominators divide in evenly, so 12 is the least common denominator.

To finish the problem, you'll have to multiply both the numerator and denominator of each fraction by the appropriate number. (Remember, that number is the result of dividing the overall least common denominator by the individual fraction's denominator):

$$\frac{1 \times 6}{2 \times 6} = \frac{6}{12}$$
$$\frac{1 \times 4}{3 \times 4} = \frac{4}{12}$$
$$\frac{3 \times 3}{4 \times 3} = \frac{9}{12}$$

You've Got Problems

Problem 2: Rewrite the following fractions so that they contain the least common denominator:

$$\frac{1}{3}, \frac{5}{6}, \frac{7}{10}$$

Operations with Fractions

Now that you have a basic knowledge about what fractions are and how you can manipulate them, it's time to start combining them using the four basic arithmetic operations.

Adding and Subtracting Fractions

In case you're wondering why I spent so much time discussing common denominators, you're about to find out. (Isn't this exciting?) It turns out that *you can only add or subtract fractions if they have common denominators.* If I had a quarter for every time a student of mine saw the problem $\frac{2}{3}+\frac{9}{11}$ and gave an answer of $\frac{11}{14}$ (they just added the numerators and denominators together), I might not be a rich man, but I could spend most of a three-day weekend at the mall playing pinball.

Here's the correct way to add or subtract fractions:

1. Write the fractions with a common denominator. (If adding a fraction to an integer, write the integer as a fraction by dividing it by 1 before getting common denominators.)
2. Add only the numerators of the fractions together, and write the result over the common denominator.
3. Simplify the fraction if necessary.

The hardest part of the entire process is getting that common denominator, and since you already know how to do that, this fraction thing is no sweat!

Example 3: Combine the fractions as indicated and provide your answer in simplest form.

(a) $\frac{3}{4}-\frac{8}{3}$

Critical Point

The answer to 3(a) can be written three different ways, all of which are correct: $-\frac{23}{12}$, $\frac{-23}{12}$, or $\frac{23}{-12}$. The same goes for any negative fraction; all of those forms are equivalent.

Solution: Even though one of these fractions is improper, you don't do anything differently. Start by rewriting these fractions with their least common denominator of 12.

$$\frac{3\times3}{4\times3}-\frac{8\times4}{3\times4}$$

$$\frac{9}{12}-\frac{32}{12}$$

Now subtract the numerators and write the result over the common denominator.

$$-\frac{23}{12}$$

Since 23 is a prime number, it won't have any factors in common with 12, so there's no way to simplify the fraction further, and you're finished.

(b) $2+\frac{4}{5}+\frac{7}{10}$

Solution: Start by rewriting 2 as a fraction. (All integers have an invisible denominator of 1.)

$$\frac{2}{1}+\frac{4}{5}+\frac{7}{10}$$

Rewrite the fractions with the least common denominator, 10.

$$\frac{2\times10}{1\times10}+\frac{4\times2}{5\times2}+\frac{7}{10}$$

$$\frac{20}{10}+\frac{8}{10}+\frac{7}{10}$$

Add the numerators together and divide the answer by the common denominator.

$$20 + 8 + 7 = 35$$

$$\frac{35}{10}$$

To finish, simplify the fraction. (Divide both numbers by the common factor of 5).

$$\frac{7}{2}$$

You've Got Problems

Problem 3: Simplify and write your answer in simplified form.

$$\frac{3}{2}-\frac{1}{3}+\frac{1}{12}$$

Multiplying Fractions

While it's a shame you can't add two fractions by just summing up their respective numerators and denominators, it's a nice surprise that in fraction multiplication, things are just that easy. You read right! All you have to do is multiply all of the numerators together and write the result divided by all of the denominators multiplied together.

Example 4: Multiply the fractions and provide the answer in simplest form.

$$\frac{1}{2}\times\frac{3}{5}\times\frac{8}{9}$$

Solution: Multiply all of the numerators, then do the same for the denominators.

$$1 \times 3 \times 8 = 24 \qquad 2 \times 5 \times 9 = 90$$

Your answer will be the numerator result divided by the denominator result. Make sure to simplify using the greatest common factor of 6.

$$\frac{24}{90}=\frac{4}{15}$$

You've Got Problems

Problem 4: Multiply and provide the answer in simplest form.

$$\frac{5}{6}\times 8\times\frac{3}{20}$$

Dividing Fractions

In Chapter 1, you learned that every number has a reciprocal. At that time, I told you the reciprocal was just 1 divided by the number. For example, the reciprocal of 8 is $\frac{1}{8}$. What if, however, you want to take the reciprocal of a fraction? Is the reciprocal of $\frac{3}{5}$ equal to $\frac{1}{\frac{3}{5}}$? That hideous-looking thing is called a *complex fraction*, and I'll talk more about that way ahead in Chapter 17.

For now, this is what you should remember: *The reciprocal of a fraction is formed by reversing the parts of the fraction* (the numerator becomes the denominator and vice versa). For example, the reciprocal of $\frac{3}{5}$ is equal to $\frac{5}{3}$, the fraction flipped upside down. (It's easy to remember what the word "reciprocal" means; just remember "re*flip*rocal.")

Wonder why I'm dredging up reciprocals again? Here's why: Dividing by a fraction is exactly the same thing as multiplying by its reciprocal. Therefore, dividing a number by $\frac{3}{5}$ gives the same result as multiplying by $\frac{5}{3}$.

Example 5: Write the answer in simplest form: $-\frac{3}{4}\div\frac{5}{16}$

Solution: Don't be psyched out by the negative sign; even though they're fractions, they're still just numbers, and the same rules you used in Chapter 1 still apply. (Since the fractions have different signs, the result will be negative.) Meanwhile, to solve the problem, take the reciprocal of the number you're dividing by and change the division sign to multiplication.

$$-\frac{3}{4}\times\frac{16}{5}$$

All that's left is a measly old multiplication problem, and you already know how to do that.

$$-\frac{3\times 16}{4\times 5}=-\frac{48}{20}=-\frac{12}{5}$$

Before I wrap up this chapter on fractions, let me hit you with an example problem that will truly see if you have mastered fractions or not. It'll test a couple of your skills at once.

You've Got Problems

Problem 5: Write the result as a fraction in simplest form.

$$\left(\frac{1}{2}-\frac{2}{7}\right)\div\frac{1}{7}$$

The Least You Need to Know

- The top of the fraction is the numerator and the bottom is the denominator.
- If a fraction is in simplest form, its numerator and denominator have no common factors.
- In order to add or subtract fractions, they must all possess a common denominator.
- The reciprocal of a fraction is simply the fraction written upside down.
- Dividing by a fraction is the same as multiplying by its reciprocal.

Chapter 3

Encountering Expressions

In This Chapter

- Representing numbers with variables
- Simplifying more complex mathematical expressions
- Unleashing the power of exponents
- Performing operations in the correct order

The fateful hour has come. Tension hangs thickly in the air like a big, fat seagull. The world, as one, holds its breath as its population understands, though only subconsciously, that something big is about to happen. Little children look toward the horizon expectantly, trying to catch a glimpse of what is to come. Dogs bark without cease, straining at their leashes to run, catching the scent of revolution in the air. Nerds everywhere stop memorizing lines from *Monty Python* movies for a moment, sensing mathematical evolution.

How's that for dramatic buildup? Perhaps it's a little overstated, but everything you used to know about math is going to change, so I thought it fitting. From this point forward, every chapter is going to contain a little less of the familiar, and will be further removed from the friendly and familiar world of numbers and arithmetic. It's time to get algebratized.

Take a deep breath, and then wade with me out into the cool but shallow waters of algebraic expressions. We'll spend a little time here during this chapter, until you get used to the water temperature. Before long, though, you'll be swimming in water too deep to stand up in, and you'll wonder why you never thought you'd be able to do it.

Introducing Variables

A *variable* is a letter that represents a number; it works sort of like a pronoun does in English. Rather than saying "Dave is a funny guy, because Dave can do an uncanny impression of Dave's grandmother," it's much more natural to say "Dave is a funny guy, because *he* can do an uncanny impression of *his* grandmother." Those pronouns *he* and *his* both clearly refer to Dave, so there's no confusion on the part of the reader, and the second sentence sounds a whole lot more natural.

Remember the directions I gave for finding the least common denominator in Chapter 2? I called the largest original denominator *Bubba*, which is actually a variable! (Although in algebra we usually use letters rather than Southern-sounding nicknames.) Once I defined *Bubba*, I could have shortened my explanation using the full power of variables like so: "I will call the largest denominator *Bubba* and the other denominators a and b (it does not matter which is which). If both a and b divide evenly into *Bubba*, then *Bubba* is your least common denominator. If not, calculate 2 × *Bubba*, and see if a and b divide evenly into that. If that number doesn't work, try 3 × *Bubba*, then 4 × *Bubba*, and so on until you find a number into which a and b divide evenly."

Critical Point

In this book (like in all math textbooks) variables are italicized so that they stand out from the surrounding text.

See how much easier it is to say "2 × *Bubba*" rather than "two times the largest denominator, as you identified in the previous step?" Besides simplifying things, however, variables serve another important purpose: They allow you to speak very precisely. Precision is important to mathematicians, who want no confusion whatsoever in their explanations.

For example, in Chapter 2 I described a fraction's reciprocal as the fraction flipped upside down. A more precise explanation could be given using variables: "The reciprocal of the fraction $\frac{a}{b}$ (such that a and b are integers) is the fraction $\frac{b}{a}$." There's no need for me to say "the old numerator becomes the new denominator," because you can see the old numerator, a, actually become the new denominator.

As useful as variables are, some math instructors overuse them. While they may provide shorter and more concise definitions, they are also more confusing to understand, especially for new algebra students. Therefore, I will almost always accompany formulas, variables, and definitions with a plain English explanation, so that you can both understand better and begin to piece together how to speak mathematically as well.

Translating Words into Math

It's no big surprise that variables act like pronouns, because (believe it or not) math is basically its own language, defined by numeric and logical rules. Therefore, one of the first skills you must master is how to translate regular English into this new (much geekier) math language. For now, I will help you translate English phrases into mathematical phrases, called *expressions*.

The expressions we'll be translating are very straightforward. For example, the phrase "3 more than an unknown number" becomes the mathematical expression "$x + 3$." Because the unknown number has no explicitly stated value, we label it with a variable x. To get a value three more than x, simply add 3 to it. If this is tricky to you, just think in terms of a real number example. Ask yourself "What is 3 more than the number 7?" Clearly the answer is 10, and you get that answer via the expression "$7 + 3 = 10$." So, to get 3 more than an unknown number, replace the known number 7 with the unknown number x.

In this translation problem, the key words were "more than," because they clued you in that addition will be necessary. Each operation has its own cue words, and I've listed them here, with an example for each:

Addition

- *More than/greater than:* "11 greater than a number" means "$x + 11$"
- *Sum:* "The sum of a number and 6" means "$x + 6$"

Subtraction

- *Fewer than/less than:* "7 fewer than a number" means "$x - 7$"
- *Less:* "17 less a number" means "$17 - x$"
- *Difference:* "The difference of a number and 6" means "$x - 6$"

Multiplication

- *Product:* "The product of a number and 3" means "$x \cdot 3$", also written "$3x$" (since multiplication is commutative, the order in which you write the numbers doesn't matter)
- *Of:* "Half of 20" means "$\frac{1}{2} \cdot 20$"

Kelley's Cautions

Since subtraction is not commutative, be careful to get the order of the numbers right. Notice that "5 less a number" and "5 less than a number" mean completely different things.

Critical Point

If no symbol is written between two things, multiplication is implied. Thus, $5y$ means "5 times y" and $2(x + 1)$ means "2 times the whole expression $(x + 1)$." Also, from this point forward, I will usually use the symbol "·" to indicate multiplication, because it's too easy to mix up the other multiplication symbol, ×, with the variable x.

Division

- *Quotient:* "The quotient of 10 and a number" means "$10 \div x$"
- *Fractions:* Don't forget that any expression written as a fraction is technically a division problem in disguise.

Example 1: Translate into mathematical expressions.

(a) The sum of 6 and twice a number

Solution: The phrase "twice a number" translates into two times the number, or $2x$. (Think about it—twice the number 8 would be $2 \cdot 8 = 16$.) The word "sum" indicates that you should add 6 and $2x$ together to get a final answer of $6 + 2x$.

(b) The product of 2 and 3 more than a number

Solution: You may be tempted to give an answer of $2 \cdot x + 3$, but that equals $2x + 3$, or "3 more than 2 times a number." You need to use parentheses to keep the $x + 3$ together like so: $2(x + 3)$. Now, the 2 is multiplied by the entire $x + 3$, not just the x.

You've Got Problems

Problem 1: Translate into a mathematical expression: 5 less than one third of a number.

Behold the Power of Exponents

You may have seen teeny little numbers floating above and to the right of other numbers and variables, like in the expression x^3. What is that little 3 doing up there? Parasailing? Is it small because of scale, perhaps because it is actually as large as the earth's sun in reality, but is only seen from hundreds of thousands of miles away? No, that little guy is called the *exponent* or *power* of x in the expression, and indeed it has a very *power*ful role in algebra.

Talk the Talk

In the exponential expression y^4, 4 is the **exponent** and y is the **base**.

Big Things Come in Small Packages

The role of an exponent is to save you time and to clean up the way expressions are written. Basically, an exponent is a shorthand way to indicate repeated multiplication.

In the language of algebra, x^3 (read "x to the third power") means "x multiplied by itself three times", or $x \cdot x \cdot x$. To find the value of real numbers raised to exponents, just multiply the large number attached to the exponent (called the *base*) by itself the indicated number of times.

Example 2: Evaluate the exponential expressions.

(a) 4^3

Solution: In this expression, 4 is the base and 3 is the exponent. To find the answer, multiply 4 by itself 3 times:

$$4 \cdot 4 \cdot 4 = 16 \cdot 4 = 64$$

Therefore, $4^3 = 64$.

(b) $(-2)^5$

Solution: In this case, the base is –2, so it should be multiplied by itself 5 times. Don't stress out about the negative signs. Just go left to right, and multiply two numbers at a time. Start with $(-2) \cdot (-2)$ to get 4 and then multiply that result by the next –2, and that result by the next –2 until you're finished.

Critical Point

Two exponents have special names. Anything raised to the second power is said to be *squared* (5^2 can be read "5 squared"), and anything to the third power is said to be *cubed* (x^3 can be read "x cubed").

$$(-2)(-2)(-2)(-2)(-2) = 4(-2)(-2)(-2) = -8(-2)(-2) = 16(-2) = -32$$

You've Got Problems

Problem 2: Evaluate the expression: $(-3)^4$.

Exponential Rules

Once you write something in exponential form, there are very specific rules you must follow to simplify expressions. Here are the five most important rules, each with a brief explanation:

- **Rule 1:** $x^a \cdot x^b = x^{a+b}$. If exponential expressions with the same base are multiplied, the result is the common base raised to the *sum* of the powers.

$$x^4 \cdot x^7 = x^{4+7} = x^{11} \qquad (2^2)(2^3) = 2^{2+3} = 2^5$$

- **Rule 2:** $\frac{x^a}{x^b} = x^{a-b}$. If you are dividing exponential expressions with the same base, the result is the common base raised to the *difference* of the two powers.

$$\frac{z^7}{z^4} = z^{7-4} = z^3 \qquad \frac{(-5)^{10}}{(-5)^9} = (-5)^1 = -5$$

- **Rule 3:** $(x^a)^b = x^{a \cdot b}$. If an exponential expression is itself raised to a power, multiply the exponents together. This is different from Rule 1, because here there is one base raised to two powers, and in Rule 1, there were two bases raised to two powers.

$$(3^5)^6 = 3^{5 \cdot 6} = 3^{30} \qquad (k^2)^0 = k^{2 \cdot 0} = k^0 = 1$$

- **Rule 4:** $(xy)^a = x^a y^a$ **and** $\left(\frac{x}{y}\right)^a = \frac{x^a}{y^a}$. If a product (multiplication problem) or quotient (division problem) or any such combination is raised to a power, then so is every individual piece within.

$$(5y)^2 = 5^2 \cdot y^2 = 25y^2 \qquad (x^2y^3)^4 = (x^2)^4 \cdot (y^3)^4 = x^8y^{12}$$

Critical Point

Any number raised to the 1 power equals the original number ($x^1 = x$); so, if there's no power written, it's understood to be 1 ($7 = 7^1$). In addition, anything (except 0) raised to the 0 power equals 1 ($x^0 = 1$, $12^0 = 1$). The expression 0^0 works a little differently, but you don't deal with that until calculus.

- **Rule 5:** $x^{-a} = \frac{1}{x^a}$ **and** $\frac{1}{x^{-a}} = x^a$. If something is raised to a negative power, move it to the other part of the fraction (if it's in the numerator, send it to the denominator and vice versa) and change the exponent to its opposite. If the expression contains other positive exponents, leave them alone.

  Most teachers consider answers containing negative exponents unsimplified, so make sure to eliminate negative exponents from your final answer. Also, note that raising something to the –1 power is equivalent to taking its reciprocal.

$$\frac{x^{-3}y^2}{z^3} = \frac{y^2}{x^3z^3} \qquad \left(\frac{4^2}{w^5}\right)^{-1} = \frac{4^{2(-1)}}{w^{5(-1)}} = \frac{4^{-2}}{w^{-5}} = \frac{w^5}{4^2} = \frac{w^5}{16}$$

Most of the time, you'll have to apply multiple rules during the same problem, in your attempts to simplify.

Example 3: Simplify the expression $\frac{(x^2y^{-3})^2}{(xy^2)^4}$.

Solution: Start by applying Rules 3 and 4 to the numerator and denominator.

$$\frac{x^{2\cdot 2}y^{-3\cdot 2}}{x^{1\cdot 4}y^{2\cdot 4}} = \frac{x^4y^{-6}}{x^4y^8}$$

Now apply Rule 2, since you have matching bases in the numerator and denominator.

$$(x^{4-4})(y^{-6-8}) = x^0y^{-14} = 1 \cdot y^{-14} = y^{-14}$$

Finish by applying Rule 5.

$$y^{-14} = \frac{1}{y^{14}}$$

You've Got Problems

Problem 3: Simplify the expression $(x^3y)^5 \cdot (x^{-2}y^2)^3$.

Living Large with Scientific Notation

One very practical application of exponents is their use in *scientific notation*, a method used to express numbers that are extremely large or extremely small.

Have you ever owned a Rubik's Cube? It's a little puzzle with six differently colored sides. According to the manufacturer, there are 43 quintillion different possible moves (all of which I have probably tried, unsuccessfully, to solve it). In case you don't know how many a quintillion is (I had to check it myself), here's the number: 43,000,000,000,000,000,000. On the other hand, here's what it looks like in scientific form: 4.3×10^{19}; even though it's weird-looking, it's a heck of a lot shorter.

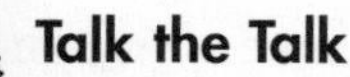

Talk the Talk

You can use **scientific notation** to express numbers that are extremely large or extremely small. The end result of scientific notation is a decimal times the number 10, which is raised to an exponent that's an integer.

In order to express a number in scientific notation, here's what you do:

- **Large numbers:** Move the decimal place in the original number to the left, counting each digit you pass, until only one nonzero digit remains to the left of the decimal place. Write the smaller decimal (ignoring any 0's at the end) and add this to the end of it: "$\times 10^n$", where n is the number of digits you passed.

  In the Rubik's example, there is no explicit decimal point, so imagine it falls at the very end of the number. You want to move it left until only the 4 remains to its left. To do so, you'll have to pass 19 numbers, hence its notation of 4.3×10^{19}.

- **Small numbers:** In these cases, you'll have to move the decimal point to the right until exactly one nonzero digit appears to the left of the equal sign. Write the resulting decimal (without any 0s before or after it) and add this on: "$\times 10^{-n}$"; n is once again the number of digits you passed.

Example 4: Write in scientific notation: .0000000000372.

Solution: Since this is a very small number, move the decimal to the right until there's exactly one nonzero digit (in this case 3) to the left of the decimal point: 3.72. In order to get that, the decimal must pass 11 digits (10 0s and the 3). Therefore, your final answer is 3.72×10^{-11}. Don't forget that the exponent is negative for small numbers.

You've Got Problems

Problem 4: Write in scientific notation:

(a) 23,451,000,000,000

(b) .00000000125

Dastardly Distribution

Many times throughout the rest of the book, you're going to see a number or variable multiplied by a quantity in parentheses, like this: $5(x + 1)$, which is read "5 times the quantity x plus 1." The parentheses are used to indicate that the 5 is multiplied by the *entire* quantity $x + 1$, not just the x nor just the 1.

According to an algebraic property (that I kept secret until now … surprise!) called the *distributive property*, you can rewrite that expression as $5x + 5$. In other words, you can multiply the 5 times everything inside those parentheses.

Talk the Talk

The **distributive property** is defined mathematically like this: $a(b + c) = ab + ac$. Notice that the a gets multiplied by the contents of the parentheses, which are then dropped; it works no matter how many items fall within the parentheses.

I think of the x and the 1 as my brother and I at Christmas time, and the 5 as a present from my grandmother. She knew that if she got us different presents, there would be strife in the house. Both of us would love our present but secretly covet the other guy's gift as well. To avoid this, my grandmother would give us both the same thing. That's how the distributive property works; it gives the same thing (in the form of multiplication) to everything in the parentheses to avoid conflict and, possibly, brawling beneath the mistletoe.

Critical Point

Occasionally, you'll see an expression like $2x - (3x - 5)$. Even though it's not written as such, this means the same as $2x - 1(3x - 5)$; the negative sign is actually an implied -1 coefficient. Therefore, to simplify the expression, you should distribute the -1 like this:

$$2x - 1(3x) - 1(-5)$$
$$= 2x - 3x + 5 = -x + 5$$

Example 5: Apply the distributive property: $-2x(3x^2 + 5y - 7)$.

Solution: Multiply everything in the parentheses by $-2x$, and add the results together:

$$(-2x)(3x^2) + (-2x)(5y) + (-2x)(-7)$$

Uh-oh … how do you multiply $-2x$ by $3x^2$? It's easy—just multiply the number parts together ($-2 \cdot 3 = -6$) and the variable parts as well, using exponential rules ($x \cdot x^2 = x^{1+2} = x^3$) to get $-6x^3$. Do the same thing with the other two products as well.

Critical Point

I'll go into greater detail about how to multiply numbers and variables together in the beginning of Chapter 10. If you're having trouble multiplying things together while using the distributive property, skip ahead and read the first couple of sections of Chapter 10, and then return here.

$$-6x^3 - 10xy + 14x$$

Notice that there's no handy way to multiply x and y together in the second term; they aren't exponential expressions with the *same* base, so you can't add the powers or anything—just leave them be and they won't hurt anyone, I promise.

You've Got Problems

Problem 5: Apply the distributive property:

$$6y^3(2x - 5y^2 + 8)$$

Get Your Operations in Order

If I asked you to evaluate the expression $4 + 5 \cdot 2$, what answer would you give? Most new algebra students answer 18, but the correct answer is actually 14. Math isn't like reading—you don't always perform the operations in an expression left to right. Instead, the order you take depends upon the operations themselves. For example, you should always perform multiplication before addition, so in the expression $4 + 5 \cdot 2$, you should multiply to get 10 and *then* add the 4, which is where the correct answer of 14 comes from. Here's the exact order you should follow when evaluating expressions:

1. **Parentheses (or other grouping symbols):** If there's more than one set of grouping symbols, work from the inside out, just like you did in Chapter 1.
2. **Exponents:** Evaluate any powers in the expression.
3. **Multiplication and Division:** These operations get done in the same step; neither one takes precedence. If there are multiple examples of one or either operation, work from left to right.
4. **Addition and Subtraction:** Again, these operations are done in the same step; if necessary, work left to right just like you did with multiplication and division.

You may have heard the mnemonic phrase "**P**lease **e**xcuse **m**y **d**ear **A**unt **S**ally" used to help remember the order of operations. Each letter in the phrase matches the first letter in the correct order of operations. However, it confuses come students, who see "m" come before "d" in the phrase and assume that multiplication always comes before division, and that's not true—they're done in the same step from left to right, as I mentioned. If you keep that in mind, however, it's a very handy phrase. If the Aunt Sally phrase is too outdated or lame for you, try "**P**oetic **e**lephants **m**ight **d**ream **a**bout **s**onnets" or "**P**enguins **e**at **m**y **d**ad's **a**thletic **s**ocks."

Example 6: Simplify the expressions.

(a) $1 + 10 \div (4 - 2)$

Solution: Parentheses come first, so subtract 2 from 4.

$$1 + 10 \div 2$$

All that's left is addition and division, and according to the order of operations, division comes first.

$$1 + 5 = 6$$

(b) $6 - [(12 + 3) \div 5 + 1]$

Solution: There are two sets of grouping symbols, so start with the innermost set, the parentheses, first.

$$6 - [15 \div 5 + 1]$$

You've still got brackets left, so simplify them, making sure to do division before addition. (The order of operations still applies inside those brackets.)

$$6 - [3 + 1] = 6 - 4 = 2$$

(c) -3^2

Solution: Let me warn you—this is kind of tricky, because most people assume that -3^2 is the same as $(-3)^2$, and it's not. The expression $(-3)^2$ translates into "–3 to the second power" or $(-3)(-3)$, which equals a positive 9.

The other expression, -3^2, means "the opposite of 3 squared." Since $3^2 = 9$, then $-3^2 = -9$.

Kelley's Cautions

Example 5(c) teaches an important lesson. In the expression $-a^n$, only a is raised to the n power, but in $(-a)^n$ it's $-a$ that's raised to the power.

You've Got Problems

Problem 6: Simplify the expression $10^2 \div 5^2 \cdot 3$.

Evaluating Expressions

I've used the word "evaluate" a lot so far, and as you've probably already figured out from context clues, it's just a fancy word for "give me the numeric answer."

Example 7: Evaluate the expression $-2x^2y$ if $x = 4$ and $y = -3$.

Solution: Since you know the values of x and y, replace (substitute) the variables with the corresponding number. I usually use parentheses to surround numbers I substitute in, to make sure I don't screw up the signs.

$$-2(4)^2(-3)$$

Now, just apply the order of operations.

$$-2(16)(-3) = (-32)(-3) = 96$$

You've Got Problems

Problem 7: Evaluate the expression $x(y - 3)^2$ if $x = 5$ and $y = -1$.

The Least You Need to Know

- Exponents are used to indicate repeated multiplication, and there are specific rules for simplifying exponential expressions.
- Scientific notation is a shorter way to write very large or very small numbers.
- The distributive property allows you to multiply a quantity times everything within a set of grouping symbols.
- You must follow a specific order of operations when simplifying any algebraic expression.

Part 2

Equations and Inequalities

Lines are nothing new to you. You've drawn them, stood in them, forgotten them on stage, and even enjoyed them on corduroy pants. In this part, however, you're going to get to know lines better than you ever have before. It'll be like getting stuck in an elevator with a neighbor you barely know. It'll be uncomfortable at first, but after a while you'll start to get used to one another, and you'll learn all sorts of interesting things about each other.

Chapter 4

Solving Basic Equations

In This Chapter

- Calculating solutions to equations
- Rules for manipulating equations
- Isolating variables within formulas
- Solving equations containing absolute values

If algebraic expressions are the logical equivalents of sentence fragments, then equations are the equivalents of full sentences. What's the difference between a phrase and a full sentence? The presence of a verb, and in the case of equations, the mathematical verb will be "equals."

Believe it or not, I think you'll find equations refreshing, especially compared with the stuff from the first few chapters. Not only was that review, but it had one major weakness: There was no easy way to tell if you got a given problem right! Sure, if asked to divide two fractions, you can probably remember to take a reciprocal and multiply, but if you make an arithmetic mistake, you're doomed! You might have had the right idea, and knew what you were supposed to do, but because of a simple screwup (like multiplying 3 by 7 and getting 10), your test paper will be covered in so much red ink that it will look like it was attacked by a badger.

That's not true for equations. When you finish solving an equation, you can test your answer, and instantly find out if you were right or not. If something went wrong, you can go back and fix it before it gets marked wrong, leaving the badger to do nothing but make whatever sounds badgers make when they're frustrated.

Maintaining a Balance

In case you haven't seen one, here's a basic equation:

$$3x - 2 = 19$$

It looks a lot like an expression, except it contains an equal sign. Just like expressions, equations translate easily into words. The above equation means "2 less than 3 times some number is equal to 19." Your job will be to figure out exactly what that number is. In this case, $x = 7$, and you can probably figure that out by plugging in a bunch of different things for x until something works. However, guessing is never a great way to reach a solution, so I'll show you a better way.

Think of an equation as a playground seesaw. The left and right sides of the equation sit on opposite sides of the seesaw, and because they are equal, it is exactly balanced, like in Figure 4.1.

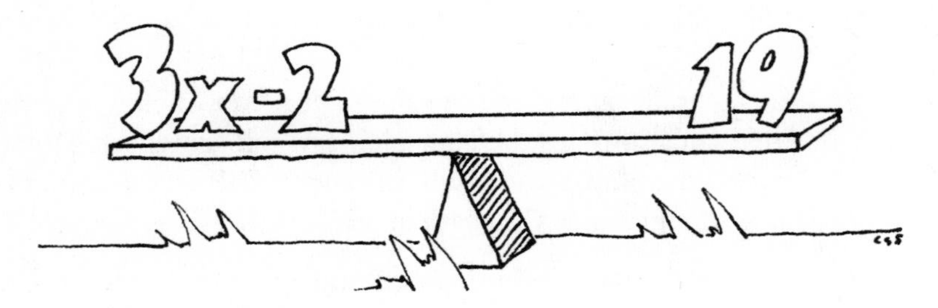

Figure 4.1

Since the two sides of the equation are equal, it's as though their weights precisely balance them on a seesaw.

> **Talk the Talk**
>
> The process of forcibly isolating a variable on one side of the equal sign (usually the left side) is called **solving** for that variable.

Your job when solving equations will be to shift around the contents of the seesaw so that only x remains in the left seat. This process is called *solving for x*, and it forces the answer to appear, as if by magic, in the right seat. You just have to be careful to keep that delicate balance the whole time.

Adding and Subtracting

Take a look at the equation $x - 5 = 11$. It's basically asking "What number, if you subtract 5, gives you 11?" You can probably arrive at the answer using only common sense, but let me show you the official mathematical way to get there.

Remember, your final goal is to get x all by itself on the left side of the equation, and the only thing in the way is –5. I want to get rid of that –5, but how? Well, if I add the opposite of –5, that will effectively cancel it out (remember the additive inverse property from Chapter 1?). However, if I suddenly add 5 to the left side of the equation, the seesaw will no longer be balanced, and someone's going to come crashing to the ground, possibly with a nasty-looking nosebleed.

To keep the seesaw balanced, I'll have to add 5 to the right side of the seesaw as well:

$$\begin{array}{rcrcr} x & - & 5 & = & 11 \\ & + & 5 & & +5 \\ \hline x & & & = & 16 \end{array}$$

Since the sum of –5 and 5 is 0, you get $x + 0$ (which equals x) on the left side of the equation. Because only x remains on the left side of the equation, the answer appears on the right side, 16.

To see if you're right, go back to the original equation, and replace x with 16:

$$x - 5 = 11$$

$$(16) - 5 = 11$$

$$11 = 11$$

Since you get a true statement (11 definitely equals 11), your answer $x = 16$ was correct.

How'd You Do That?

By replacing x with 16 to check your answer, you're applying the substitution property (another property to add to your list), which allows you to substitute values for one another, as long as they are equal.

You've Got Problems

Problem 1: Solve the equation for x.

$$8 + x = 19$$

Multiplying and Dividing

In your endeavors to isolate x's to solve equations, you'll often have to eliminate not only numbers added to or subtracted from x, but numbers actually *attached* to the x, like in the equation $5x = 45$. To be precise, the 5 and x are not actually glued to one another. They are actually just multiplied, and 5 is called the *coefficient*. Therefore the equation is actually asking "5 times what number is equal to 45?"

Talk the Talk

When a number is written next to a variable, that number is called a **coefficient**. Thus, in the expression $2x$, 2 is the coefficient. Note that coefficients are usually written at the front of the expression, before the variable.

To answer that question, you can't add or subtract 5 like in the previous example, because neither addition nor subtraction can cancel out multiplication. You've got two options when it comes to eliminating coefficients; notice that both of the techniques require the same thing be done to both sides of the equation, just like before—you've got to keep that seesaw balanced!

- **Integer coefficients:** If the x has a coefficient of n, divide both sides by n to get your final answer.
- **Fractional coefficients:** If the x has a rational coefficient $\frac{a}{b}$, multiply both sides of the equation by $\frac{b}{a}$ to solve for x. This may look familiar, because it's based on the multiplicative inverse property.

Both of these techniques change the coefficient of x into some fraction $\frac{c}{c}$; in other words, x's coefficient will now have the same number in both the numerator and denominator, and any such fraction is equal to 1. (Remember, any nonzero number divided by itself equals 1.) Therefore, the left side of the equation can be rewritten as $1x$, which means the exact same thing as plain old x, so your isolation procedure is complete.

Example 1: Solve the equations and check your answers.

(a) $5x = 45$

Solution: Since the coefficient is an integer, divide both sides of the equation by it.

$$\frac{5x}{5} = \frac{45}{5}$$
$$x = 9$$

To check the answer of 9, plug it in for x in the original equation.

$$5(9) = 45$$

$$45 = 45$$

(b) $\frac{2}{3}y = 12$

Solution: Since the coefficient is rational, multiply both sides of the equation by its reciprocal, $\frac{3}{2}$.

$$\frac{3}{2} \cdot \frac{2}{3}y = \frac{3}{2} \cdot \frac{12}{1}$$
$$\frac{6}{6}y = \frac{36}{2}$$
$$y = 18$$

To check the answer, substitute 18 for y in the original equation.

$$\frac{2}{3}\left(\frac{18}{1}\right) = 12$$
$$\frac{36}{3} = 12$$
$$12 = 12$$

You've Got Problems

Problem 2: Solve the equation.

$$-\frac{4}{5}w = 16$$

Equations with Multiple Steps

Most of the time, solving equations requires more than a single step. For example, think about the equation I introduced you to at the beginning of the chapter: $3x - 2 = 19$. Not only is there a -2 on the same side of the equal sign as the x, but there's also a 3 clinging to that x, like a dryer sheet stuck to a pant leg. In order to isolate x (and therefore solve the equation), you'll have to get rid of both numbers, using each of the techniques you've learned so far. (It wouldn't hurt to throw in some fabric softener with static cling controller as well.)

If a solution requires more than a single step, here's the order you should follow:

1. **Simplify the sides of the equation separately.** Each of the items added to or subtracted from one another in the equation are called *terms.* If two terms have the exact same variable portion, then they are called *like terms*, and you can combine them as though they were numbers.

Talk the Talk

Consider the equation $3y - 7y = 12$. Since $3y$ and $-7y$ both have the exact same variable part (y), they are called **like terms** and you can simplify by combining the coefficients and leaving the variable alone: $3y - 7y = -4y$, since $3 - 7 = -4$, so the equation is now $-4y = 12$.

I'll carefully define *like terms* and discuss them further in Chapter 10.

2. **Separate the variable.** Using addition and subtraction, move all terms containing the variable you're isolating to one side of the equation (usually the left) and move everything else to the other side (usually the right). You're finished when you have something that looks like this: $ax = b$ (a number times the variable is equal to a number).
3. **Eliminate the coefficient.** If the variable's coefficient is something other than 1, you need to either divide by it or multiply by its reciprocal (like you did earlier in this chapter).

Equation solving requires practice, and it's going to take some trial and error before you get good at it. Don't forget to check your answers! Even though I will only show answer checking rarely from this point forward (to save space), rest assured that I never let the chance to make sure I got the answer right go by! Eventually, you'll feel comfortable checking answers by substituting in your head and working things out mentally.

Example 2: Solve each equation.

(a) $3x - 2 = 19$

Solution: You can't simplify the left side, since $3x$ and -2 are not like terms, so the first thing to do is to separate the variable term. Accomplish this by adding 2 to both sides.

$$\begin{array}{rcrcr} 3x & - & 2 & = & 19 \\ & + & 2 & & +\ 2 \\ \hline 3x & & & = & 21 \end{array}$$

Divide both sides by 3 to eliminate the coefficient.

$$\frac{3x}{3} = \frac{21}{3}$$
$$x = 7$$

(b) $-14 = 2x + 4(x + 1)$

Solution: You can do a bit of simplifying on the right side of the equation. Start by distributing that positive 4 into the quantity within parentheses.

$$-14 = 2x + 4 \cdot x + 4 \cdot 1$$

$$-14 = 2x + 4x + 4$$

Simplify like terms $2x$ and $4x$.

$$-14 = 6x + 4$$

At this point, the problem looks a lot like the equation from part (a), except the variable term appears on the right side of the equation. There's no problem with that—it's perfectly fine. In fact, if you leave the $6x$ on the right side, it's less work to separate the variable term. Just subtract 4 from both sides.

$$\begin{array}{rcrcr} -14 & = & 6x & + & 4 \\ -4 & & & - & 4 \\ \hline -18 & = & 6x & & \end{array}$$

Divide both sides by 6 to eliminate the coefficient.

$$\frac{-18}{6} = \frac{6x}{6}$$

$$-3 = x$$

How'd You Do That?

In Example 2, part (b), I solved the equation by isolating the x on the right side, rather than the left side. To tell you the truth, I prefer x on the left side as a matter of personal taste, even though it doesn't affect the answer at all.

According to the *symmetric property* of algebra, you can swap sides of an equation without affecting its solution or outcome. In other words, I could have flip-flopped the sides of the equation in 2(b) to get $2x + 4(x + 1) = -14$. If you solve that equation, you'll get $x = -3$, the exact same answer. So, if you ever wish the sides of an equation were reversed, go ahead and flip them without fear.

(c) $-3(x + 7) = -2(x - 1) + 5$

Solution: You can apply the distributive property on both sides of the equal sign to begin.

$$-3x - 21 = -2x + 2 + 5$$

Simplify the right side by combining the 2 and 5 (which are technically like terms, since they have the exact same variable part—no variables at all).

$$-3x - 21 = -2x + 7$$

Now, it's time to separate the variable term. Do this by adding $2x$ to both sides (to remove all x terms from the right side of the equation) and adding 21 to both sides as well (to remove plain old numbers from the left side of the equation).

$$\begin{array}{rcrcrcr} -3x & - & 21 & = & -2x & + & 7 \\ +2x & + & 21 & & +2x & + & 21 \\ \hline -x & & & = & & & 28 \end{array}$$

Critical Point

As demonstrated in Example 2(c), a negative variable like $-w$ technically has an implied coefficient of -1, so you can rewrite it as $-1w$ if you wish. (This is similar to implied exponents, where a plain old variable like w has an implied coefficient of 1, so $w = 1w^1$.)

At this point, you have $-x = 28$, which means "the opposite of the answer equals 28." Therefore, the correct answer is $x = -28$ (since -28 is the opposite of 28).

Here's another way to get the final answer: Since $-x = -1 \cdot x$, you can rewrite the final line of the equation so it looks like it has a coefficient and divide by that -1 coefficient:

$$\frac{-1x}{-1} = \frac{28}{-1}$$
$$x = -28$$

(d) $y + 3 = \frac{1}{4}y + 5$

Solution: Since there are no like terms together on one side of the equation, skip right to separating the variable terms. Accomplish this by subtracting $\frac{1}{4}y$ and 3 on both sides of the equation. (By the way, even though the variable is y, not x, in this equation, that doesn't change the way you solve it.)

$$\begin{array}{rcrcrcr} y & + & 3 & = & \frac{1}{4}y & + & 5 \\ -\frac{1}{4}y & - & 3 & & -\frac{1}{4}y & - & 3 \\ \hline \frac{3}{4}y & & & = & & & 2 \end{array}$$

Since the coefficient is fractional, multiply both sides by its reciprocal to finish.

$$\frac{4}{3}\left(\frac{3}{4}y\right)=\left(\frac{4}{3}\right)\frac{2}{1}$$
$$\frac{12}{12}y=\frac{8}{3}$$
$$y=\frac{8}{3}$$

Since 8 and 3 have no common factors (other than 1), the improper fraction cannot be simplified, so leave it as is.

You've Got Problems

Problem 3: Solve the equations.

(a) $3(2x - 1) = 14$

(b) $2x - 7 = 4x + 13$

Absolute Value Equations

If you're trying to solve an equation containing a variable that's trapped inside absolute value bars, you'll need to slightly modify your technique. Here's why: Such equations may, in fact, have two answers, instead of just one! That might be initially shocking and exciting (like finding out that while you thought you had one secret admirer, you actually have two), once you figure it all out in the end, everything makes sense. (It must be because you are one irresistible hunk of burnin' love.)

Here's what to do if you encounter an equation whose poor, defenseless variable is trapped in absolute value bars:

1. **Isolate the absolute value expression.** Just like you isolated the variables before, this time isolate that entire expression that falls between the absolute value bars. Follow the same steps as before—start by adding or subtracting things out of the way and finish by eliminating a coefficient, if the expression has one.
2. **Create two new equations.** Here's the tricky part. You're actually going to design two completely separate equations from the original one. The first equation should look just like the original, just without the bars on it. The second should look just like the first, only take the opposite of the right side of the equation. This might sound tricky, but trust me, it's easy.

Critical Point

You're creating two separate equations because absolute values change two different values (any number and its opposite) into the same thing. If you don't understand what I mean, check out the end of

3. **Solve the new equations to get your answer(s).** Both of the solutions you get are answers to the original absolute value equation.

To remind myself that absolute value equations require two separate parts, I sometimes imagine that those absolute value bars are little bars of dynamite that blow that original equation in half, creating two distinct pieces. Now that you get the idea, let me show you how to handle the explosives correctly.

Example 3: Solve the equation $4|2x-3|+1=21$.

Solution: Start by isolating the absolute value quantity on the left. To do so, first subtract 1 from both sides.

$$4|2x-3|=20$$

To complete the isolation process, divide both sides by 4.

$$\frac{4|2x-3|}{4}=\frac{20}{4}$$
$$|2x-3|=5$$

Now that only the absolute values remain on the left side, it's time to create two new equations. The first looks just like the above equation (without bars attached); its sister equation is an exact replica, except its right side will be the opposite of its sibling's right side (it'll have –5 instead of 5).

$$2x-3=5 \qquad\qquad 2x-3=-5$$

Solve those equations separately.

$$\begin{aligned} 2x-3&=5 \\ 2x&=8 \\ x&=4 \end{aligned} \qquad\qquad \begin{aligned} 2x-3&=-5 \\ 2x&=-2 \\ x&=-1 \end{aligned}$$

There you go; the answers are –1 and 4. Do you find two answers hard to swallow? Watch what happens when I check them both in the original equation.

$$\begin{aligned} 4|2(4)-3|+1&=21 \\ 4|5|+1&=21 \\ 4(5)+1&=21 \\ 20+1&=21 \end{aligned} \qquad\qquad \begin{aligned} 4|2(-1)-3|+1&=21 \\ 4|-5|+1&=21 \\ 4(5)+1&=21 \\ 20+1&=21 \end{aligned}$$

Notice that the contents of the absolute values are opposites, so once the absolute value is taken, you end up getting the same results.

You've Got Problems

Problem 4: Solve the equation $|x-5|-6=4$.

Equations with Multiple Variables

So far, when I've asked you to solve an equation for a variable, it was pretty obvious which one I was talking about. For example, to solve the equation $3x + 2 = 23$, you'd solve for (isolate) the x variable. Why? Because it's the only variable in there! That x is enjoying all the attention, like the only girl in an all-boys school.

I need to add another skill to your equation-solving repertoire that will be extremely important in Chapter 5: how to solve for a variable when there's more than one variable in the equation. Don't worry—you don't need to learn a whole new set of steps to follow or anything; it's basically done the same way you're used to. Here's the only difference: When you're finished, you'll have variables on both sides of the equation, rather than just one side.

Example 4: Solve the equation $-2(x-1) + 4y = 5$ for y.

Solution: Start by simplifying the left side of the equation.

$$-2x + 2 + 4y = 5$$

Now it's time to separate the variable term. Since you're trying to isolate the y, eliminate every term not containing a y from the left side of the equation. In other words, add $2x$ to, and subtract 2 from, both sides of the equation.

$$4y = 2x + 3$$

It might have felt weird to move that $-2x$ term, but it had to go, since you're solving for y, not x. All that's left to do now is eliminate y's coefficient by dividing both sides by 4.

$$y = \frac{2x+3}{4}$$

Even though that answer is correct, it will make more sense in Chapter 5 if you rewrite it slightly. Any time two or more things are added or subtracted in the numerator of a fraction, you can break that fraction into smaller fractions, each of which will contain one term of the original numerator and a copy of the denominator. (If the terms contain variables, just stick them to the right of the fraction.)

$$y = \frac{2}{4}x + \frac{3}{4}$$

$$y = \frac{1}{2}x + \frac{3}{4}$$

You've Got Problems

Problem 5: Solve the equation $9x + 3y = 5$ for y.

The Least You Need to Know

- In order to solve equations correctly, you must keep them balanced at all times—always perform the same operations on both sides of the equation simultaneously.
- To solve an equation "for" a variable means to isolate that variable on one side of the equal sign.
- You can always check your solutions to equations by substituting them in for the variable in the original equation.
- If a variable appears inside absolute value signs within an equation, you may get two solutions, because you have to rewrite the equation as two different equations to solve it.

Chapter 5

Graphing Linear Equations

In This Chapter

- Navigating the coordinate plane
- Plotting points and graphing lines
- Identifying the slope of a line
- Graphing linear absolute values

The fact that I am a horrible artist is a running joke with me. Although I am pleased with my understanding of math, I would give my left arm for the ability to draw something as meaningless as a realistic-looking chicken. (My feeble attempts at a chicken scratch would probably end up looking something like a polar bear balanced on top of a four-slot restaurant toaster.)

Luckily, mathematical drawings are much more technical (read: boring) than creative drawings. In this chapter, I will focus on drawing mathematical graphs—pictures of things like equations, but not things like arctic animals or food service appliances. In fact, by the end of this chapter, you'll be able to draw pictures of linear equations, whose graphs are plain old straight lines. Nothing's easier to draw than a straight line, right? (Even though my attempts to draw straight lines rather look like drawings of chickens.)

Climb Aboard the Coordinate Plane

Thanks to a fun-loving mathematician named René Descartes (a guy who contributed so much to math that I won't tease him for having a girl's name), we have the vastly useful mathematical tool called the *coordinate plane*. It is basically a big, flat grid used to visualize mathematical graphs. Check out Figure 5.1 for a look at the coordinate plane.

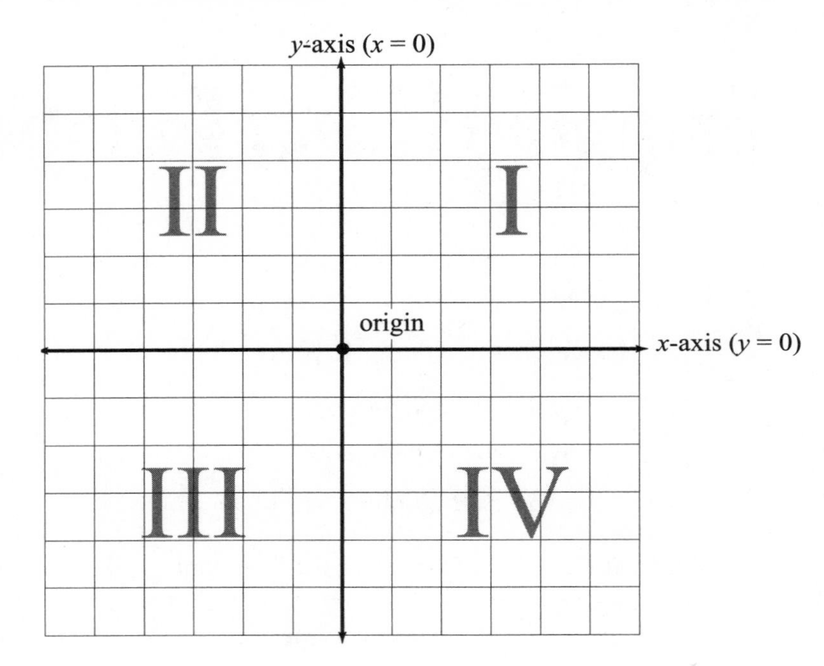

Figure 5.1

The horizontal x-axis *and the vertical* y-axis *intersect at the origin, and divide the plane into four quadrants (labeled with Roman numerals).*

Talk the Talk

The **coordinate plane** is a flat grid used to visualize mathematical graphs. It is formed by the **x-axis** and **y-axis,** which are, respectively, horizontal and vertical lines that meet at a point called the **origin**. The axes split the plane into four **quadrants**.

To best understand the coordinate plane, pretend that it is the map of a small town. In this town, there are only two main roads, one that runs horizontally (called the *x-axis*) and one that runs vertically (called the *y-axis*). These two roads intersect once, right in the middle of town, at a location called the *origin*. This intersection splits the town into four *quadrants*, which are numbered in a very specific way. The northeast section of town is quadrant one (I), the northwest is quadrant two (II), the southwest is quadrant three (III), and the southeast is quadrant four (IV).

Just like normal roads, the x- and y-axes also have more official-sounding addresses, in addition to their names. The address of the x-axis is the equation $y = 0$. In fact, every one of the horizontal roads in this town has the address "$y =$ *something*." The first horizontal road above the x-axis in Figure 5.1, for example, has equation (address) $y = 1$, the road above that has equation $y = 2$, and so on. The first horizontal road *below* the x-axis has equation $y = -1$, the next is $y = -2$, etc. Basically, all of the positive horizontal street addresses occur in quadrants I and II, and the negative ones are located in quadrants III and IV.

In a similar way, the vertical streets (all of which have equations "$x =$ *something*") are numbered as well. The y-axis has equation $x = 0$, and the streets to its right begin with $x = 1$, and get larger and larger. On the other hand, the street immediately to the left of the y-axis has equation $x = -1$, and the further left you go, the more and more negative the streets get. So, positive vertical streets run through quadrants I and IV, and the negative vertical streets run through quadrants II and III.

Critical Point

You can scale each grid line on the coordinate plane; each line could represent 1, 2, 5, 10, or any fixed number of units. However, unless otherwise noted in this book, always assume that each grid line on the coordinate plane represents exactly 1 unit of measure.

Because all of the streets have these handy addresses, it's very simple to pinpoint any location in town. For example, let's say I find a great bakery at the intersection of $x = -3$ and $y = 4$ streets as shown in Figure 5.2.

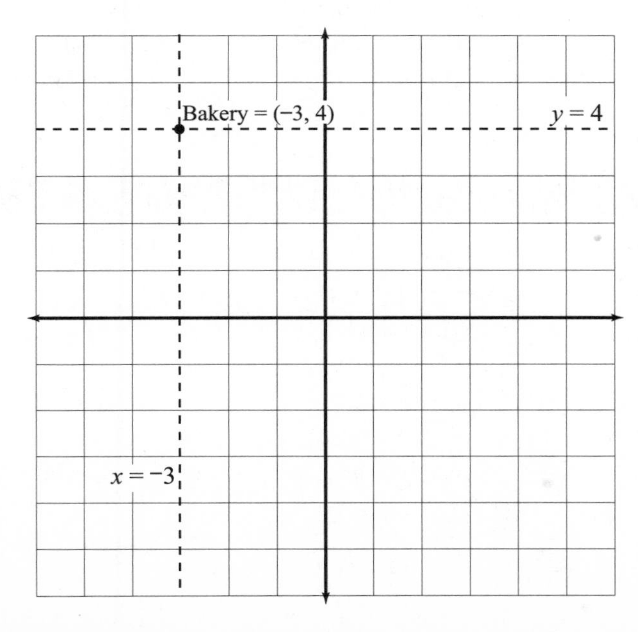

Figure 5.2

You're going to love the fresh-baked bagels in this quadrant II bakery.

Talk the Talk

Every point on the coordinate plane is described by a **coordinate pair** *(x,y)*. The *x* portion of the coordinate pair is sometimes called the **abscissa,** and the *y* portion is called the **ordinate,** but that terminology is very old, antiquated, and formal, so you may not hear it unless your teacher is old and antiquated.

Written on the door of the bakery is its official coordinate plane address: (–3, 4). You see, every location in the coordinate plane has an address (x, y) called a *coordinate pair*, based on the intersecting street numbers. When writing the coordinate pair, make sure to list the x street first, and then the y street.

I hope you weren't too insulted by the simplicity of my map metaphor for the coordinate plane, but I find it more interesting than the strict mathematical definition. Once you get good at plotting (graphing) points on the coordinate plane, you'll be ready to do more advanced things, like connecting those dots to form graphs.

Example 1: Plot these points on the coordinate plane: $A = (2,0)$, $B = (0,-4)$, $C = (-3,-2)$, $D = \left(-\frac{7}{2},3\right)$, $E = (5,-1)$, and $F = (6,2)$.

Solution: Remember that each coordinate pair represents the intersection of a vertical street (the first number in the pair) and a horizontal street (the second number in the pair). For instance, point C lies at the intersection of the third vertical street to the left of the origin and the second horizontal street below it.

Points A and B will fall on the x- and y-axes, respectively, since they each contain a 0 in the coordinate pair. The trickiest point to plot is D, since it contains a fraction. To make things easier for you, convert the improper fraction $-\frac{7}{2}$ into the mixed number $-3\frac{1}{2}$ (using the technique you learned in Chapter 2). To plot D, count three and a half units to the left of the origin and then three units up. All of the answers to Example 1 appear in Figure 5.3.

You've Got Problems

Problem 1: Identify the points indicated on the coordinate plane of Figure 5.4.

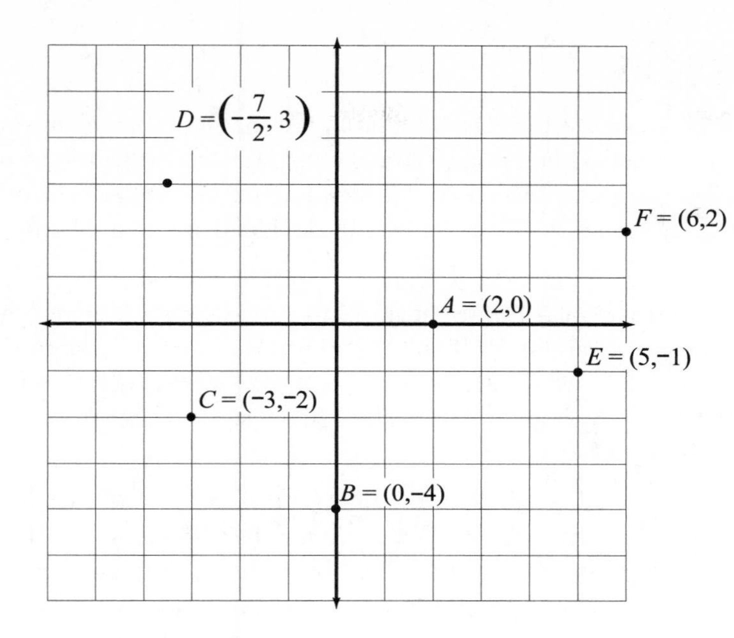

Figure 5.3

The solution to Example 1.

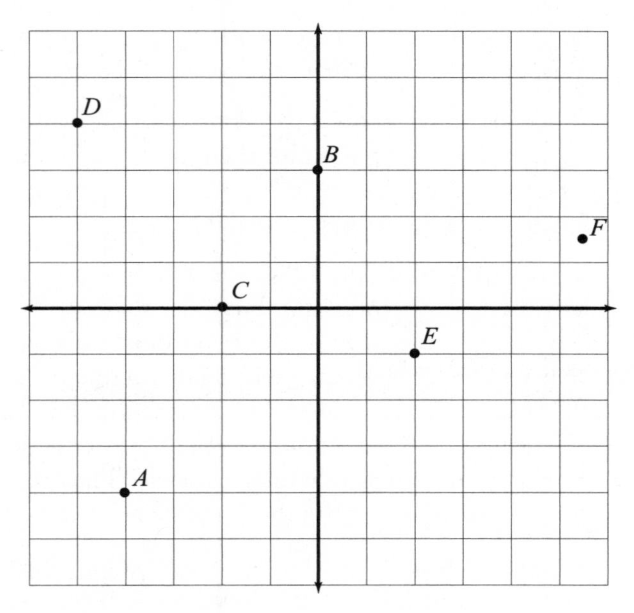

Figure 5.4

List the (x,y) *coordinates for each of the points* A *through* F.

Sketching Line Graphs

Not that plotting points isn't fun, but it gets old kind of fast. So, let's up the ante and graph some lines, which are only slightly more complicated. Lines are the graphs of *linear equations*, which are equations that are in this form: $ax + by = c$. In other words, they're equations that contain x and y (with coefficients attached) and a number without a variable, called a *constant*.

Back in Chapter 4, you solved simpler linear equations like $2x - 1 = 15$; they're simpler because they only have one variable in them. For example, it's pretty easy to figure out that the solution to the equation $2x - 1 = 15$ is $x = 8$. However, when there are two variables present in an equation, a nutty thing happens—now there's not just one solution, but an *infinite number of solutions!*

Talk the Talk

A **linear equation** has the form $ax + by = c$, where x and y are variables, a and b are their coefficients, and c is a plain old number with no variable, called a **constant.** Here are some examples of linear equations: $x - y = 5$, $3x - 2y = 1$, and $4y - 2 = x$. (The terms don't always have to be in the same order.)

How'd You Do That?

Is $x = 7$ a linear equation? Yes! I know I said a linear equation has the form $ax + by = c$, and it's true that the equation $x = 7$ doesn't have any y's in it, so how can it be linear? You can skirt this technicality by saying that $b = 0$; in other words, the coefficient of y is 0, so there is a y, but there's no use writing it.

Take a look at the linear equation $x + y = 9$, which translates to "What two numbers add up to 9?" There are many pairs of numbers x and y that could make that statement true. If $x = 1$ and $y = 8$, you get 9. How about $x = 11$ and $y = -2$? Those two numbers add up to 9 as well! How can you possibly give a solution to this equation when there are so many dang answers?

The best thing to do is to write each set of solutions as an ordered pair and plot them on a graph. For example, given the two solutions I just came up with, I can create the points (1,8) and (11,–2).

Here's the cool thing: All of the solution points, not just the two I pointed out, but *all* of them, will make a perfectly straight line in the coordinate plane, and that line (since geometry tells us that a line is made up of an infinite number of points) represents the infinite number of solutions to the linear equation. Finally, a math word that makes sense. They're called *linear* equations because their graphs are *lines*.

If you ever wondered why you had to graph equations, now you know! The graph is just a visual representation of all the ordered pairs (x,y) that, if plugged back into the linear equation, would make it true. Now that you know *why* you have to make graphs, I'll focus on *how* to make them.

Graphing with Tables

The easiest way to graph any equation in algebra is to use a table. Basically, you'll plug a whole bunch of numbers in for x and find out what the corresponding y values are to complete the ordered pair. Once you do, you can plot that (x,y) pair on the coordinate plane, secure in the knowledge that it falls on the graph.

Here are the steps to follow to graph a linear equation using a table:

1. **Solve the equation for *y*.** This makes simplification easier later, and since I just showed you how to do this at the end of last chapter, why not show off your skill?
2. **Plug in a few values for *x* and record the resulting *y* values.** Write each corresponding x and y pair together as a coordinate pair (x,y).
3. **Plot the points on the coordinate plane and connect them to form the graph.** In these early stages of graphing, I suggest you use graph paper. Later on, as you get more experienced, you'll be able to sketch rough graphs without needing as much precision.

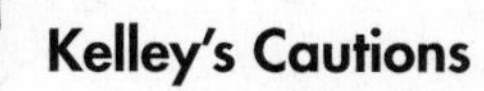

Kelley's Cautions

How many x-values should you plug in? Well, geometry tells us that it takes two points to define a line, but if you plot three, you can use the third one to check yourself. If they don't all fall on the same line, then you made an algebra mistake.

Example 2: Sketch the graph of $2x - y = 5$ using a table.

Solution: Solve for y by subtracting $2x$ from both sides and then either multiplying or dividing both sides by –1 to make y positive.

$$-y = -2x + 5$$
$$y = 2x - 5$$

Now it's time to construct the table. The left column will contain some x values to be plugged in. (I usually choose –1, 0, and 1 since they are small, simple numbers.) The middle column is used to perform the calculations, and the right column is used to record the resulting coordinate pair.

x	$y = 2x - 5$	(x, y)
-1	$y = 2(-1) - 5$ $y = -2 - 5 = -7$	$(-1, -7)$
0	$y = 2(0) - 5$ $y = -5$	$(0, -5)$
1	$y = 2(1) - 5$ $y = 2 - 5 = -3$	$(1, -3)$

Now that you know the points (–1,–7), (0,–5), and (1,–3) are solutions, plot them and connect the dots to get the graph (pictured in Figure 5.5).

Figure 5.5

The line 2x – y = 5 will extend infinitely in each direction. Note that the x- and y-axes do not appear in the exact middle of this graph; since all three of the points occur below the x-axis, I have shifted the focus of my graph there.

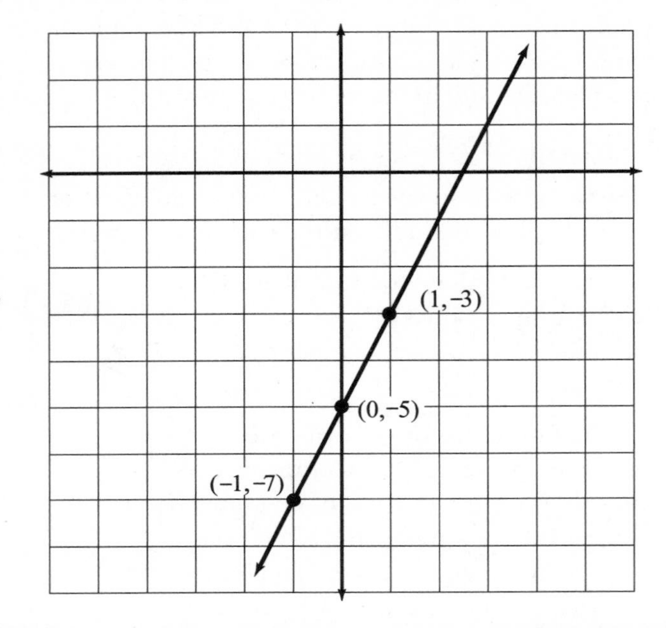

You've Got Problems

Problem 2: Sketch the graph of $-4x + y = 2$ using a table.

Graphing with Intercepts

The spot (or spots) where a graph intersects either the *x*- or *y*-axis is called an *intercept.* You may have noticed, as you practiced plotting points, that coordinates which fall on the axes will always contain a 0 in them. In fact, a coordinate on the *x*-axis will always have a *y*-value of 0, and a coordinate on the *y*-axis will always have an *x*-value of 0. This makes calculating intercepts easy. All you have to do to find an *x*-intercept is plug in 0 for *y*, and to calculate a *y*-intercept, plug in 0 for *x*.

Talk the Talk

An **intercept** is a point on either the *x*- or *y*-axis through which the graph passes. Every *x*-intercept has the form $(x,0)$, and every *y*-intercept has the form $(0,y)$.

Usually, plugging in 0 for a variable makes students happy, because anything multiplied by that 0 will vanish. So, it's not a very burdensome task. Even better, once you've found those two intercepts, you can plot them and connect the dots to find the graph of the line!

Example 3: Graph $3x - y = 6$ by calculating its intercepts.

Solution: If you plug in 0 for x and solve, you'll get the value of the y-intercept.

$$3(0) - y = 6$$
$$-y = 6$$
$$y = -6$$

The y-intercept is (0,–6). Now plug 0 in for y (back in the original equation) and solve to get the value of the x-intercept.

$$3x - (0) = 6$$
$$3x = 6$$
$$x = 2$$

The x-intercept is (2,0). Plot the two intercepts and connect the dots to get the graph, pictured in Figure 5.6.

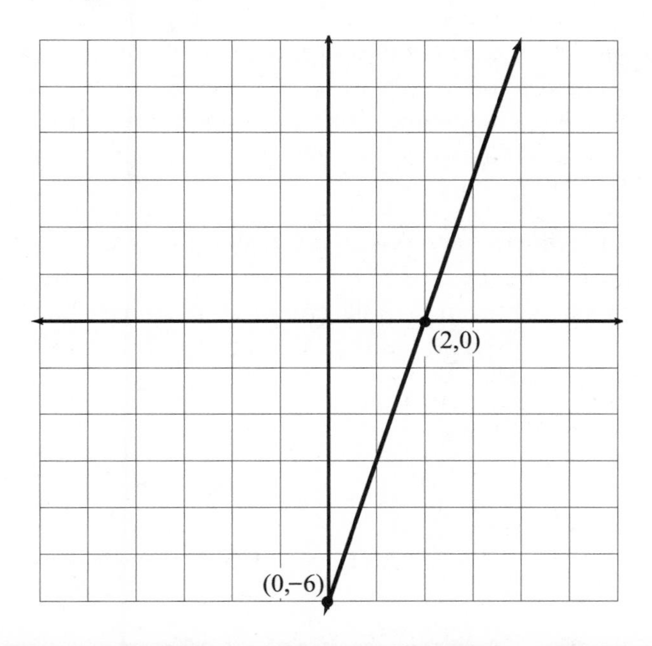

Figure 5.6

The solution to Example 3, the graph of $3x - y = 6$.

You've Got Problems

Problem 3: Graph $4x + 2y = -8$ by calculating its intercepts.

It's a Slippery Slope

The *slope* of a line is a number that describes how "slanty" that line is. Is the line steep or shallow? Does it rise from left to right or fall? Even if you don't have the graph of the equation, you can answer all of these questions knowing only the slope of the line in question.

Calculating Slope of a Line

As useful as the slope of a line is, it's remarkably easy to calculate. All you have to do is pick out two points on the line, which I'll call points A and B, for reference. The slope of the line is equal to this fraction:

$$\frac{\text{number of vertical units it takes to get from } A \text{ to } B}{\text{number of horizontal units it takes to get from } A \text{ to } B}$$

Let me explain what I mean in terms of a real-life example—placekicking in football. Whenever a kicker comes onto the field to try for an extra point or a field goal, he must make sure that his path to the ball is precise. He needs to get a running start at the ball to give his kick some momentum. (If he just stood there and swung his leg, the ball wouldn't go very far.) However, if his running path is off course, the ball won't go exactly where he's intending it.

To make sure their approach paths are exactly right, all right-footed kickers do the same thing: They stand over the ball, take two steps backward, and then two steps to their left to reach their starting position. This creates an imaginary line of approach, which passes through two points: The position of the ball and their starting position, as shown in Figure 5.7.

Figure 5.7

To position himself correctly, the kicker takes two steps back from the ball and then two steps left.

Imagine we're looking down at the field, perhaps from an airplane or a blimp advertising tires, and you can easily calculate the slope of that approach line. The slope's numerator is equal to the number of steps he takes forward or backward (up or down as we look down on the field); since he backs up two steps from the ball, the numerator will be –2. (If he'd have gone forward, and moved up according to our perspective, the number would have been positive.)

The denominator of the slope equals the number of steps he takes left or right, where left steps are negative and right steps are positive. Since he takes two steps left, the denominator equals –2. Therefore, the slope of the line is:

$$\frac{-2}{-2} = 1$$

So, you can easily calculate the slope of a line by picking two of its points and counting the number of steps (or units) it takes to get from one to the other. However, this is not a very handy technique, especially if the points' coordinates aren't integers.

Critical Point

You can also calculate the slope of the kicker's approach by going from the kicker to the ball, rather than vice versa. In that case, the numerator is 2 (two steps forward or up) and so is the denominator (two steps right). So, the slope equals $\frac{2}{2}$, which simplifies to 1. So, no matter which you consider the starting and ending point, you get the same answer.

There is a handy formula you can use to calculate the slope of a line, given the coordinates of two of its points; the formula basically counts the horizontal and vertical changes for you. It works like this: If a line contains points (a,b) and (c,d), then the slope of the line is equal to

$$\frac{d-b}{c-a}$$

In other words, subtract the y-values from one another, and divide that by the difference of the x-values.

Kelley's Cautions

Make sure to subtract the y-values in the numerator and x-values in the denominator. Students often reverse them, mistakenly placing the x's in the top of the fraction, so be cautious.

Example 4: Calculate the slope of the line that passes through the points (4,–1) and (–3,9) and explain what it means.

Solution: Apply the slope formula, which finds the difference of the y's and divides it by the difference of the x's:

$$\frac{9-(-1)}{-3-(4)}=\frac{9+1}{-3-4}=\frac{10}{-7}=-\frac{10}{7}$$

Remember, if a fraction is negative, you can rewrite the fraction, placing the negative sign in the denominator or the numerator, so $-\frac{10}{7}=\frac{-10}{7}=\frac{10}{-7}$.

What do these slopes mean? A slope of $\frac{-10}{7}$ means that you can go down 10 units and right 7 units from one point on the line to reach another point on that line, because that's exactly how to travel from (4,–1) to (–3,9). (Graph those coordinate pairs if you don't believe me.)

Alternatively, the slope $\frac{10}{-7}$ assures you that you can go up 10 units and left 7 units from one point on the line to reach another point, because that's the path from (–3,9) to (4,–1).

You've Got Problems

Problem 4: Calculate the slope of the line passing through the points (4,0) and (–5,6).

Interpreting Slope Values

You can tell a lot about a line just by looking at its slope. Here are a few facts to familiarize yourself with about the gripping, romantic relationship between a line and its slope:

- **Lines with positive slopes rise.** If a line's slope is greater than 0, its graph will slant upwards from left to right.
- **Lines with negative slopes fall.** In other words, the coordinates on the line will have smaller and smaller y values as you travel left to right.
- **Lines with large slopes are steep.** The further a slope's absolute value is from 0, the steeper its graph will be. In fact, a line with slope 2 is significantly steeper than a line with slope 1, so the steepness increases sharply and quickly as the slope values increase. A line with a slope of –5 is also very steep, it's just slanted in a different direction than lines with positive slopes.
- **Lines with small slopes rise more gradually.** The closer a slope's absolute value is to 0, the more its graph will resemble a horizontal line.
- **Horizontal lines have slopes equal to 0.** Any two points on a horizontal line will have the same y value, so when you calculate the slope, the numerator will be 0.
- **Vertical lines have undefined slopes.** Learn more about this in the nearby sidebar.

Critical Point

What does it mean when a vertical line has an *undefined* slope? Basically, the slope of a vertical line is an illegal number. Take, for example, the vertical line $x = 3$; I'll choose two random points from the line, (3,4) and (3,9), and use the slope formula:

$$\frac{9-4}{3-3}=\frac{5}{0}$$

Did you know that you're not allowed to divide by 0 in math? It's illegal! It's just as much an anathema as ripping those little mattress tags off.

Any fraction with 0 in the denominator (but not in the numerator) is said to be *undefined,* so that's why your textbook says the slope of a vertical line is undefined, or that the line has "no slope."

How can you remember that vertical slopes are undefined but horizontal slopes are 0? Think about trying to walk along the lines. How much effort would it take to walk along a horizontal line? Zero.

How much effort would it take to walk along a vertical line? You can't! It's impossible, because there is no slope in a vertical line. It'd be like trying to walk up a wall, and unless you have been bitten by a mutant spider, wear brightly colored red-and-blue spandex, and have a girlfriend named Mary Jane, that skill's probably not on your resumé.

Kinky Absolute Value Graphs

One final word about graphing linear equations before I wrap this chapter up. Remember in Chapter 4 when you learned that equations containing an x in absolute values required a slightly different solution method than regular, nonabsolute value equations? You had to break the equation into two parts to get the answer. Well, you have to graph them a little differently than regular linear equations, too. Whereas the graph of a normal linear equation looks like a line, an absolute value linear equation graph looks like a "V." It's basically a line with a kink in it, a sharp point (or *vertex*) where the graph changes directions.

In Figure 5.8, I've drawn the graphs of $y = x - 3$ and $y=|x-3|$. Both of the graphs have the same coordinates for their x values greater than 3.

However, when x is less than 3, the left graph dips below the x-axis (meaning that its y values are negative). See how the right graph makes a sharp turn in order to avoid going below the x-axis altogether? This is because the equation shown in the right graph sets y equal to an absolute value! Remember, absolute values can never be negative, and the graph reflects this by avoiding negative y values like the plague.

Figure 5.8

The graph of $y = |x - 3|$ *takes drastic measures to avoid negative values, unlike* $y = x - 3$.

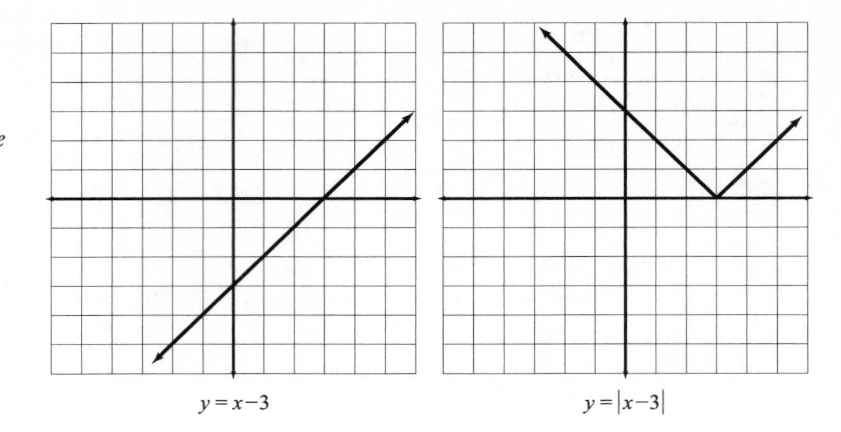

The best way to graph a linear absolute value is to find exactly where its vertex is, plot it, and then plot one point to the right and one point to the left of it, so you can draw the branches of the graph.

Example 5: Graph the equation $y = -\frac{1}{2}|x+4|-3$.

Solution: To find the vertex of the graph, set just the contents of the absolute values equal to 0 and solve for x.

$$x + 4 = 0$$
$$x = -4$$

Now find the corresponding value for y when $x = -4$ to get the coordinate pair for the vertex. (Plug $x = -4$ back into the original equation.)

$$y = -\frac{1}{2}|(-4)+4|-3$$
$$y = -\frac{1}{2}(0)-3$$
$$y = -3$$

The vertex of the graph occurs at (–4,–3), so plot that point on the coordinate plane. Now choose one x value to the left of the vertex and one to the right (in other words, choose one x which is smaller than –4 and one which is larger), and plug them both into the original equation. I chose $x = -6$ and $x = -2$.

> CAUTION
>
> **Kelley's Cautions**
>
> If you draw an absolute value graph, and it dips below the x-axis, that doesn't necessarily mean you did it wrong! Some linear absolute value equations will *still* have negative y's. For example, if you plug $x = 2$ into the equation $y = |x| - 5$, you'll get $y = -3$. I'll discuss this in greater detail in Chapter 16.

$$y = -\frac{1}{2}|(-6)+4|-3 \qquad y = -\frac{1}{2}|(-2)+4|-3$$
$$y = -\frac{1}{2}|-2|-3 \qquad y = -\frac{1}{2}|2|-3$$
$$y = -\frac{1}{2}(2)-3 \qquad y = -\frac{1}{2}(2)-3$$
$$y = -1-3 \qquad y = -1-3$$
$$y = -4 \qquad y = -4$$

Plot the resulting coordinate pairs, (–6,–4) and (–2,–4), each time drawing a line that begins at the vertex and passes through one of the points, as shown in Figure 5.9.

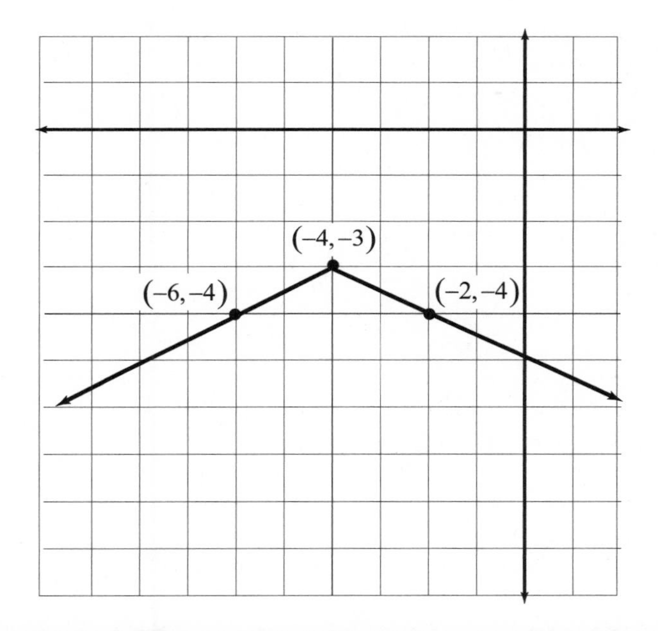

Figure 5.9

The graph of $y = -\frac{1}{2}|x+4|-3$, the solution to Example 5. Note that this coordinate plane is shifted to highlight quadrant III.

You've Got Problems

Problem 5: Graph the equation $y = |2x - 4| + 1$.

The Least You Need to Know

- The coordinate plane is a grid defined by a horizontal line called the x-axis and a vertical line called the y-axis, which meet at the origin.
- Vertical lines have equation "$x = c$," and horizontal lines have equation "$y = d$," where c and d are real numbers.
- You can graph linear equations by plotting points or calculating their intercepts.
- The slope of a line is a ratio describing how "slanty" a line is by examining how much vertical and horizontal change occurs between points on the line.
- The slope of a horizontal line is 0, and a vertical line has an undefined slope.
- The graph of a linear absolute value equation has a sharp point in it called the "vertex."

Chapter 6

Cooking Up Linear Equations

In This Chapter

- Determining equations of lines
- Applying the point-slope and slope-intercept formulas
- Writing linear equations in standard form
- Exploring connections to geometry

Now that you can graph lines, and have some idea of what a slope is, it's time to get down and dirty with linear equations. Last chapter, you were introduced to the concept, but now it's time to create your own equations from scratch. Even if you're not very good at cooking or baking, don't worry. The recipes for creating linear equations are pretty simple, require very few ingredients, and almost never result in an oven fire.

All you need to create a linear equation is its slope and one of the points on the line. Once you've got those two pieces of information, you'll be cranking out equations as fast as Sarah Lee can whip out pound cakes. Along the way, you'll even get a better understanding of linear equations and how all of their pieces fit together correctly. By the end of this chapter, you'll be a pro at lines; maybe even the wait at the Department of Motor Vehicles won't bother you anymore!

Point-Slope Form

If you are given the slope of a line and one of the points on the line, then creating the equation of that line is a very simple procedure. All you'll need is the *point-slope formula* for a line.

Point-slope formula: If a line has slope m and passes through the point (x_1,y_1), then the equation of the line is

$$y - y_1 = m(x - x_1)$$

Are you wondering where that m came from? For some reason, math people have used the variable m to represent the slope of a line for a long time. Believe it or not, no one quite knows why. I could wax historical about this mathematical conundrum, but you'd get bored fast, so let me suffice to say that m is the variable used to represent slope in all of the formulas you'll see in this chapter.

Basically, all you have to do to create a linear equation is to plug in a slope for m, an x-value from an ordered pair for x_1, and the matching y-value for y_1, and simplify.

Critical Point

Remember, no matter which of the techniques described in this chapter you use to find a linear equation, you always need two things: the slope of the line and a point on the line.

Example 1: Write the equation of the line with slope –3 that passes through the point (–1,5) and solve the equation for y.

Solution: Since the slope equals –3, set $m = -3$ in the point-slope formula. You should also replace x_1 with the known x-value (–1) and replace y_1 with the matching y-value (5).

$$y - y_1 = m(x - x_1)$$

$$y - (5) = -3(x - (-1))$$

Simplify the right side of the equation.

$$y - 5 = -3(x + 1)$$

$$y - 5 = -3x - 3$$

Since the problem asks you to solve for y, you should isolate it on the left side of the equation by adding 5 to both sides.

$$y = -3x + 2$$

Critical Point

Variables with subscripts, like x_1 and y_1, will have completely different values than their non-subscripted look-alikes, x and y. By the way, that little subscript does not affect the value of the variable at all, like an exponent would. It's just a little garnish that distinguishes between the variables, making them different.

That's all there is to it! This is the only line in the world that has slope –3 and passes through the point (–1,5).

You've Got Problems

Problem 1: Write the equation of the line with slope 4 that passes through the point (2,–7) and solve the equation for y.

Slope-Intercept Form

Are you wondering why I asked you to solve the linear equations in the above exercises for y? I wasn't just making you jump through hoops (although I am impressed by your agility, I must admit); there's actually a very good reason to do it. Once a linear equation is solved for y, it is said to be in *slope-intercept form.*

There are two big benefits that result when you put an equation in slope-intercept form (and you can probably figure out what they are based on the name): You can identify the slope and y-intercept of the line (get this) *without doing any additional work at all!* Plus, if you transform equations while you do sit-ups, the slope-intercept form could actually give you greater definition in your abs!

Let's get mathematical for a moment. Officially speaking, the slope-intercept form of a line is written like this:

$y = mx + b$, where m is the slope and b is the y-intercept

In other words, once you solve a linear equation for y, the coefficient of x will be the slope of the line, and the number with no variable attached (called the *constant*) marks the spot on the y-axis, $(0,b)$, where the line passes through.

Example 2: Identify the slope and the coordinates for the y-intercept given the linear equation $x - 4y = 12$.

Solution: Remember, all you have to do to transform an equation into slope-intercept form is to solve it for y. To isolate the y, subtract x from both sides of the equation and then divide everything by the coefficient of –4:

$$-4y = -x + 12$$

$$y = \frac{1}{4}x - 3$$

Talk the Talk

Once a linear equation is solved for y, it is in **slope-intercept form,** $y = mx + b$. The coefficient of the x term, m, is the slope of the line and the number (or **constant**), b, is the y-intercept.

The x-term's coefficient is $\frac{1}{4}$, so the slope of the line is $\frac{1}{4}$. Since the constant is –3, the graph of the equation will pass through the y-axis at the point (0,–3). (Don't forget that the x-coordinate of a point on the y-axis will always be 0, and vice versa.)

You've Got Problems

Problem 2: Identify the slope and the coordinates for the y-intercept given the linear equation $3x + 2y = 4$.

Graphing with Slope-Intercept Form

You already know a couple of ways to graph a linear equation, but I thought I'd toss one more method into the mix. I know you don't need a hundred different ways to graph lines any more than you need a hundred different ways to tie your shoes, but since graphing using the slope-intercept form of a line is my favorite technique, I really want to share it with you. It'll be a bonding moment for us. Besides, there's nothing more boring than plotting point after point to come up with a graph. This way is a little different, and it helps you understand how the slope works, in case you are a little fuzzy on that.

Here are the steps to follow in order to graph a line using slope-intercept form:

1. **Solve the equation for *y*.** The equation needs to be in slope-intercept form.
2. **Determine the slope and *y*-intercept of the line.** Remember, this is as easy as looking at the numbers on the right side of the slope-intercept equation.
3. **If the slope is negative, rewrite it with the negative sign in the numerator or denominator.** You can stick that negative sign either place (but not both at the same time), it doesn't matter which.
4. **Plot the *y*-intercept of the graph.** Remember, the y-intercept has coordinates $(0,b)$; mark that point on the graph.
5. **Start at the *y*-intercept and use the slope to count steps to another point on the graph.** You learned how to do this in Chapter 5. Remember, positive numbers in the numerator mean up, negative numbers in the numerator mean down, positives in the denominator mean right, and negatives in the denominator mean left. For example, a slope of $\frac{-2}{3}$ means to count down 2 units and right 3 units from the y-intercept to get another point on the graph.

6. **Connect the two points.** Just connect the dots to finish the graph. Remember, it extends infinitely long in both directions.

How'd You Do That?

In case you're wondering where the slope-intercept form comes from, and how we can be so sure that the *x*-coefficient is the slope and *b* is the *y*-intercept, I'll clue you in on its origins.

Let's say there's a line with slope *m* and *y*-intercept (0,*b*) (the familiar variables from slope-intercept form). Apply point-slope form to get the equation of that line.

$$y - y_1 = m(x - x_1)$$
$$y - b = m(x - 0)$$
$$y - b = mx$$

Just solve for *y* and you end up with the slope-intercept form.

$$y = mx + b$$

A star is born!

Example 3: Graph the equation $5x + 3y = 12$ using the slope-intercept form.

Solution: Start by solving the equation for *y*.

$$3y = -5x + 12$$
$$y = -\frac{5}{3}x + 4$$

The *y*-intercept is (0,4); its slope should be rewritten with the negative sign in either the numerator or denominator: $\frac{-5}{3}$ or $\frac{5}{-3}$.

Plot the *y*-intercept and count your way to the next point based on the slope you chose. You should either count down five and right three units $\left(\frac{-5}{3}\right)$ or count up five and left three units $\left(\frac{5}{-3}\right)$. Either way, when you connect the dots, you'll end up with the same line, as shown in Figure 6.1.

I sort of like this method of graphing lines because it feels like I'm reading a treasure map: Start at the big palm tree, and then take five paces due south and three paces due east to reach the Golden Booty (which, by the way, sounds like a great name for an R&B star).

Figure 6.1

Either road you choose, you still end up on the same line.

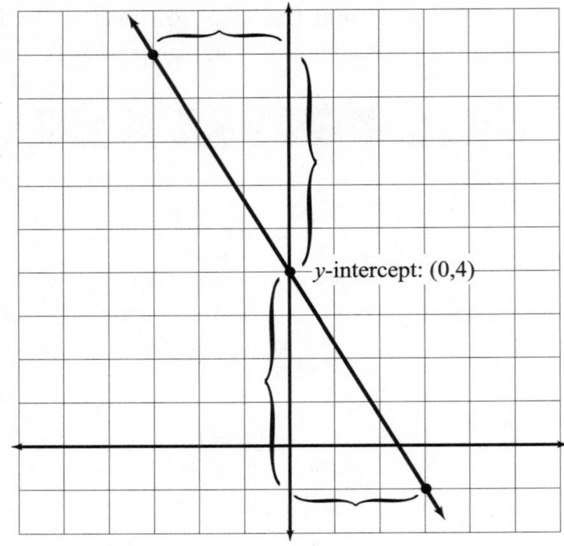
You can either count up 5 and left 3 units from the y-intercept ...
y-intercept: (0,4)
... or down 5 and right 3 units from the y-intercept to find another point on the line.

You've Got Problems

Problem 3: Graph the equation $-2x - y = 1$ using the slope-intercept form.

Standard Form of a Line

As long as you keep an equation balanced, you can do just about anything to both of its sides to change the way it looks. For example, take a look at these two linear equations:

$$3x - 2y = 4 \quad \text{and} \quad 1 + \frac{1}{2}y = \frac{3}{4}x$$

They may not look anything alike, but they are equivalent! In case you don't believe me, I'll prove it to you. I'll start with the right-hand equation and play with it until it looks exactly like the left-hand equation. To start, I'll multiply both of its sides by 4.

$$4\left(1 + \frac{1}{2}y\right) = 4 \cdot \frac{3}{4}x$$

$$4 + \frac{4}{2}y = \frac{12}{4}x$$

$$4 + 2y = 3x$$

How'd You Do That?

I usually write fractional coefficients like this: $\frac{1}{3}y$. However, you could also write the same quantity as $\frac{y}{3}$, and the value would be the same. Why? Technically, you can give the variable a denominator of 1 and multiply fractions: $\frac{1}{3} \cdot \frac{y}{1} = \frac{y \cdot 1}{3 \cdot 1} = \frac{y}{3}$.

Now, I'll flip-flop the sides of the equation, which is allowed according to the symmetric property.

$$3x = 4 + 2y$$

Finally, I'll subtract $2y$ from both sides and the transformation is complete.

$$3x - 2y = 4$$

Isn't that much better-looking than $1 + \frac{1}{2}y = \frac{3}{4}x$? Most math teachers think so. In fact, they feel so strongly about it that they usually require you to write your answers in the tamer, less-fractiony version on quizzes and tests.

This prettier version of the equation is called *standard form*, and it has the following properties:

- An equation in standard form looks like $ax + by = c$; in other words, the x and y terms are on the left side of the equation and the constant is on the right side.
- The constant and both of the coefficients must be integers.
- The coefficient of the x-term must be positive.

Why bother putting equations in standard form? To be honest, I'm not sure why teachers stress this so much. I much prefer slope-intercept form, since its values m and b actually represent something, whereas the coefficients of standard form really have no practical meaning.

Talk the Talk

An equation in **standard form** looks like $ax + by = c$, where b and c are integers, and a is a positive integer.

Some say that standard form is important since every known linear equation can be put into standard form, but vertical lines (like $x = 2$) cannot be put in slope-intercept form. (Remember, slope-intercept form means solving for y, and if there's no y around, that's impossible.) Maybe that's

Critical Point

If a linear equation is in standard form $ax + by = c$, you can use the shortcut formula $m = -\frac{a}{b}$ to find the slope. For example, the slope of $5x - 8y = -2$ will $m = -\left(\frac{5}{-8}\right) = \frac{5}{8}$

true, but in my opinion, standard form is preferred because people hate fractions—even your algebra teacher, although she'd never admit it.

Example 4: Put the linear equation $-\frac{2}{3}-4x=\frac{5}{9}y$ in standard form.

Solution: The first order of business is to get rid of all those ugly fractions. Take a look at the denominators in the equation (3 and 9) and calculate the least common denominator. (In case you forgot how to do this, review the process in Chapter 2; it had something to do with a number named *Bubba*.)

In this case, the least common denominator is 9, so multiply both sides of the equation by 9 written as a fraction $\left(\frac{9}{1}\right)$:

$$\frac{9}{1}\left(-\frac{2}{3}-\frac{4}{1}x\right)=\frac{9}{1}\left(\frac{5}{9}y\right)$$
$$-\frac{18}{3}-\frac{36}{1}x=\frac{45}{9}y$$
$$-6-36x=5y$$

Move $5y$ to the left side of the equation by subtracting it from both sides. Also add 6 to both sides, so the constant ends up on the right side of the equation.

$$-36x-5y=6$$

Remember, the x-term has to be positive in order for the equation to be in standard form, but right now it's not. No sweat—just multiply both sides of the equation by –1.

$$36x+5y=-6$$

CAUTION

Kelley's Cautions

Remember, only the x-term has to be positive in standard form, not the y-term or the constant. As you see in Example 4, changing the sign of the x-term (when necessary) will often result in one or more of the other terms becoming negative, and that's okay.

You've Got Problems

Problem 4: Put the linear equation $\frac{5}{4}y=\frac{7}{3}+\frac{1}{6}x$ in standard form.

Tricky Linear Equations

For the rest of the chapter, I want to focus on the skill of designing linear equations when you're not given an explicit slope. You're still going to use the point-slope form to write the equation, it's just that you'll have to figure out the slope first. Finding the slope won't be hard, but the method you use to find it will vary based on what sort of information is given in the problem.

Essentially, you'll be faced with two kinds of tricky linear equation problems, one in which you're given two points on the line and asked to create the equation, and one in which you have to create a line either perpendicular or parallel to a second line.

How to Get from Point *A* to Point *B*

One classic type of algebra problem asks you to write the equation of a line that passes through two points. Back in Chapter 5, you calculated the slope of such a line using something I creatively called the "slope formula." In case you forgot that formula, it said that a line passing through points (a,b) and (c,d) has the following slope:

$$m = \frac{d-b}{c-a}$$

Now it's time to put that formula to good use. Not only will you find the slope of the line passing through a couple of points, you'll find the equation of that line as well.

Example 5: Write the equation of the line that passes through the points (–4,7) and (2,–1) in standard form.

Solution: Remember, to create a linear equation, you need the slope of the line and one point on the line. You're actually given two points on the line for this problem, but you still need the slope, which is calculated with the slope formula like so:

$$m = \frac{-1-(7)}{2-(-4)}$$

$$m = \frac{-8}{6}$$

$$m = -\frac{4}{3}$$

Now that you have the slope, pick one of the two points you're given and apply the point-slope form. Here's what that substitution looks like if you choose the point (2,–1):

$$y - y_1 = m(x - x_1)$$

$$y-(-1) = -\frac{4}{3}(x-(2))$$

$$y+1 = -\frac{4}{3}x + \frac{8}{3}$$

The problem does ask you to put this in standard form, so eliminate those fractions by multiplying everything by 3.

$$3(y+1)=\frac{3}{1}\left(-\frac{4}{3}x+\frac{8}{3}\right)$$
$$3y+3=-4x+8$$

Rearrange the terms to comply with the requirements of standard form.

$$4x+3y=5$$

Critical Point

Once you figure out the slope in Example 5, it doesn't matter which point you use in point-slope form—either one will result in an equivalent equation. In case you need to see it to believe it, here's what the work looks like if you choose the point (–4,7) instead of (2,–1):

$$y-y_1=m(x-x_1)$$
$$y-(7)=-\frac{4}{3}(x-(-4))$$
$$y-7=-\frac{4}{3}x-\frac{16}{3}$$
$$3(y-7)=\frac{3}{1}\left(-\frac{4}{3}x-\frac{16}{3}\right)$$
$$3y-21=-4x-16$$
$$4x+3y=5$$

Pretty cool, eh? The final answers are the same.

Here's a tip: You can check your answer by plugging both of the original points into the equation you got. If they each result in true statements, you did everything right:

Check (–4,7)	Check (2,–1)
$4(-4)+3(7)=5$	$4(2)+3(-1)=5$
$-16+21=5$	$8-3=5$
$5=5$	$5=5$

You've Got Problems

Problem 5: Write the equation of the line that passes through the points (–3,5) and (–8,0) in standard form.

Parallel and Perpendicular Lines

You may not know a lot about geometry yet, and that's okay. (By the way, the technical definition of geometry, according to Webster's Dictionary, is "What an acorn says when it grows up." Get it? "Gee-I'm-a-tree?") However, you may be asked to construct equations of parallel and perpendicular lines in algebra, so here's a basic explanation of those concepts:

- *Parallel lines* never intersect one another. Like railroad tracks, they run on and on but never touch one another. This is due to the fact that parallel lines have *the same exact slope*, which keeps the lines the same exact distance from one another forever and ever.
- *Perpendicular lines* intersect one another at right, or 90-degree, angles. If that doesn't make sense, think of it this way: If planted in a flat plot of land, a tree will grow perpendicular to the ground. Perpendicular lines have slopes which are *opposite reciprocals* of one another.

Talk the Talk

Parallel lines do not intersect because they have the same slope. However, **perpendicular lines** meet at 90-degree angles, thanks to slopes which are opposite reciprocals. In other words, if lines *l* and *n* are perpendicular and the slope of line *l* is $\frac{a}{b}$, then the slope of *n* would be $-\frac{b}{a}$.

In Figure 6.2, I've graphed a set of parallel lines and a set of perpendicular lines and indicated their slopes. In case you're wondering where I got the slope values, I have also highlighted points on the lines; if you want to generate the slopes for yourself, just plug their coordinates into the slope formula.

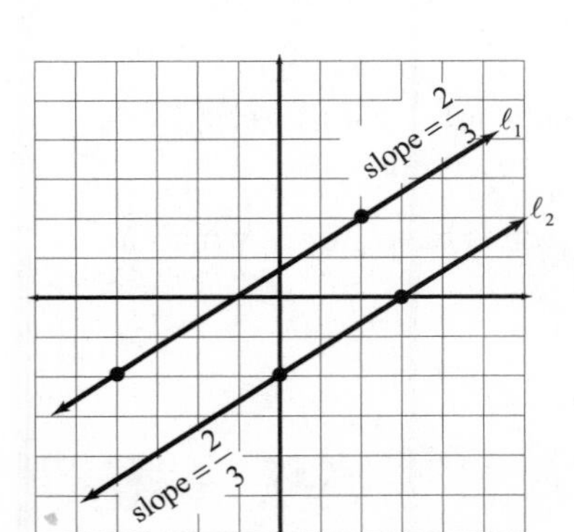

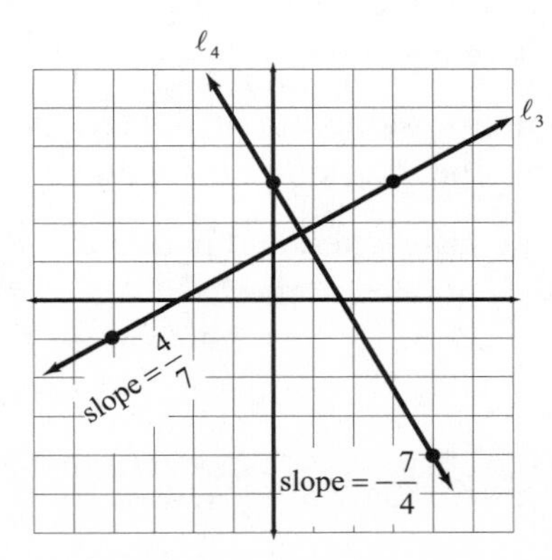

Figure 6.2

Lines l_1 and l_2 are parallel, because they have the same slope. On the other hand, the slopes of the perpendicular lines are opposite reciprocals of one another.

Example 6: Write the equation of line k in slope-intercept form if it passes through the point (2,–3) and is perpendicular to the line with equation $x - 5y = 7$.

Solution: You already know a point on k, so basically all you need is the slope of k in order to create its equation. You do know that k will be perpendicular to the line $x - 5y = 7$, so put that line in slope-intercept form to figure out what its slope is. In other words, solve $x - 5y = 7$ for y:

$$y = \frac{1}{5}x - \frac{7}{5}$$

This line has slope $\frac{1}{5}$. Therefore, line k's slope will be the opposite reciprocal of $\frac{1}{5}$, which is $-\frac{5}{1} = -5$; just flip the fraction upside down and multiply it by –1.

Now that you have the slope of k and one of its points, apply the point-slope form.

$$y - (-3) = -5(x - (2))$$
$$y + 3 = -5x + 10$$

Since the problem asks for the answer in slope-intercept form, solve for y.

$$y = -5x + 7$$

You've Got Problems

Problem 6: Write the equation of line j in standard form if j passes through the point (–6,1) and is parallel to the line $-2x + 6y = 7$.

The Least You Need to Know

- The point-slope formula says that a line with slope m which passes through the point (x_1, y_1) will have equation $y - y_1 = m(x - x_1)$.
- To put an equation into slope-intercept form, solve it for y: $y = mx + b$. The slope of the line will be m, and b is the equation's y-intercept.
- Parallel lines have equal slopes, and perpendicular lines have slopes which are opposite reciprocals of one another.

Chapter 7

Linear Inequalities

In This Chapter

- Differentiating between equations and inequalities
- Solving and graphing basic inequalities
- Interpreting compound inequalities
- Graphical solutions for linear inequalities

When Thomas Jefferson wrote that "all men are created equal" in the Declaration of Independence, he probably meant it in a philosophical and political way. That is to say, all people should have the same rights and responsibilities as citizens of a country. Practically speaking, it would be hard to argue that all men (and women) are *actually* created equal. If that were true, it wouldn't make much difference if the coach of the St. Louis Rams decided to let me play quarterback during the playoffs.

If all men were created equal, I would possess immeasurable athletic talent, and could throw a 60-yard pass that resulted in the winning touchdown, to the praise and glory of thousands of fans, who would chant my name and buy my bobble-head dolls at the concession stand. In real life, if I were in for a single play, we both know what would happen. I'd get sacked by the meanest 350-pound defensive linemen you've ever seen, instantly breaking every major bone in my body, and single-handedly destroying any chance that my face would ever be attached to any toy, whether its head bobbled or not. (Unfortunately, my real head would probably bobble permanently from that point on, though the effect is less comical in real life, I assure you.)

In any case, people are rarely created exactly equal, and so, too, mathematical statements are often unequal. In this chapter, I'll describe algebraic sentences, called *inequalities*, that aren't equations because their sides are unequal. The good news is that solving such statements involves a process nearly identical to solving equations. The only major difference you'll find is that the graphs of inequalities are a bit different than the graphs of equations.

Equations vs. Inequalities

Since the two sides of an inequality are not equal, there's got to be a reason why. The obvious explanation is that if the sides don't have the same value, then one of them must be larger; in fact, knowing which side is larger is very important. Therefore, you'll use one of four inequality symbols (in place of an equal sign) when writing an inequality statement. These symbols fulfill the important responsibility of identifying the larger side, and also tell you whether or not there is even a remote possibility that the sides could ever be equivalent:

- $<$: This symbol means "less than." Use this to indicate that the quantity on the left side has a smaller value than the quantity on the right side. For example, the statement $-2 < 7$ literally means "negative 2 is less than 7."
- $>$: If you flip the less than symbol around, you get the "greater than" symbol, which means the exact opposite of the $<$ symbol. Therefore, you could write $100 > 57$.
- $\leq$: This means "less than or equal to"; basically, it means the same as the $<$ symbol, except that it also allows both sides to have the same value. In other words, the statement $6 \leq 10$ is true (since 6 is less than 10), but so is $7 \leq 7$ (since 7 is equal to itself).

Critical Point

There is a fifth inequality symbol, $\neq$, which means "not equal to." For example, you could correctly state that $10 \neq 4$ (10 is not equal to 4), which is true but not very helpful. It would be better to say $10 > 4$ (10 is greater than 4), because that statement explains *why* the quantities are unequal.

Kelley's Cautions

When it comes to negative numbers, the *larger* a negative a number is, the *less* it is considered (which almost seems backward). Therefore, $-17 < -9$ since both numbers are negative and 17 is larger than 9. For the same reason, you can write $-4 > -9$.

- $\geq$: The "greater than or equal to" symbol is similar to the less than or equal to symbol, in that it also allows for the possibility of equality. Therefore, the statements $-5 \geq -12$ and $-15 \geq -15$ are both true.

Critical Point

Here's one way to remember which inequality sign is which: The *less* than symbol points *left*. Just remember "less goes left."

Example 1: Are the following statements true or false?

(a) $-7 > -1$

Solution: False; since 7 is larger than 1, –7 is less than –1.

(b) $1 \geq 1$

Solution: True; 1 is either greater than *or* equal to 1 (only one of those two conditions can be true at a time, and it's true that 1 is equal to 1).

You've Got Problems

Problem 1: Are the following statements true or false?

(a) $-3 < -3$

(b) $5 \leq 11$

Solving Basic Inequalities

Most of the inequalities you'll see in algebra will contain variables; although a statement like $1 < 7$ is easy enough to prove correct, it's really not all that intellectually stimulating or interesting. Instead, you'll be asked to solve an inequality statement like $-5x + 3 > -32$. Basically, your job will be to find the values of x that make that statement true. It's pretty much the same objective you had when solving equations, except in the world of inequalities, there are lots of solutions, not just one.

The Inequality Mood Swing

To solve an inequality that contains only one variable, just follow the same steps you used to solve equations. In other words, simplify both sides of the inequality, isolate the variable, and then eliminate the variable's coefficient. However, there is one major difference between equations and inequalities: When solving an inequality, if you ever multiply or divide both sides by a negative number, you must reverse the inequality sign.

What do I mean by "reverse" the inequality sign? Change less than signs to greater than signs, and vice versa. (Less than or equal to signs become greater than or equal to signs, and vice versa.) I call this the inequality mood swing. Remember, it only happens when you multiply or divide by a negative number, and that only occurs when you're trying to eliminate the coefficient. So, just remember to check for a negative coefficient as you're eliminating it, and reverse the inequality sign as necessary.

Critical Point

When you solved equations, you were advised to do the same things to both sides of the equal sign. Follow the same steps now, but do everything to both sides of the inequality sign instead.

Example 2: Solve the inequality $-5x + 3 > -32$.

Solution: Since both sides are already simplified (no sides contain like terms), isolate the variable by subtracting 3 on both sides of the greater than sign. Notice that the inequality sign does not change yet, because you're not *multiplying* or *dividing* by a negative number.

$$-5x > -35$$

Time to eliminate the coefficient. Do so by dividing both sides by -5. Don't forget to reverse the inequality sign since you're dividing by a negative number.

$$x < 7$$

So, any number less than 7, when plugged in for x, should make this inequality statement true. You obviously can't check them all to make sure your answer is right (you'd spend the rest of your life checking your work on one problem), but it doesn't hurt to check one answer just to make sure you're not way off. Here's how you could check the value $x = 6$, which should work, since it's less than 7:

$$-5(6) + 3 > -32$$
$$-30 + 3 > -32$$
$$-27 > -32$$

Since 32 is larger than 27, then -32 is definitely less than -27, so $x = 6$ is a valid answer.

You've Got Problems

Problem 2: Solve the inequality $2(w - 6) \leq 18$.

Graphing Solutions

In Chapter 5, when I was discussing linear equations, I explained why it was important to draw graphs. Since linear equations contained two variables, usually x and y, there were an infinite number of ordered pairs that made each equation true. Because you couldn't write that infinite list of answers, it was useful to represent them with a drawing (graph) of the solutions.

You now know that basic inequalities have an infinite number of solutions, so you should also use graphs to help visualize their solutions. However, basic inequalities (like those you just learned to solve in the previous section) are different than linear equations because they only contain one variable. Therefore, you don't need a coordinate plane to graph basic inequalities; all you need is a *number line*, pictured in Figure 7.1; essentially, the number line is just the x-axis from the coordinate plane; since there is no second variable, you don't need a second axis on the graph.

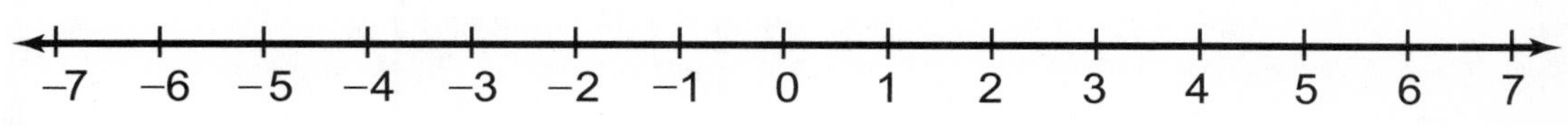

Figure 7.1

The number line can be used to graph inequalities that contain only one unique variable.

Here's how to graph the solution to a basic inequality:

1. **Solve the inequality.** Before you can draw the graph of an inequality, you need to know its solution.
2. **Draw a number line.** The number line doesn't always have to be centered at 0. It's sometimes more useful to center the number line at or near the value you found in the solution.
3. **Dot the hot spot.** When you solve the inequality, you'll get something like this: $x \leq a$ or $x > a$, where a is a real number. If the inequality symbol allows for the possibility of equality ($\leq$ or $\geq$), draw a solid dot at a. If, on the other hand, the symbol is either $<$ or $>$, draw an "open dot," or a circle, at the a value.

CAUTION

Kelley's Cautions

Make sure to get x on the left side of the inequality when you solve it. If you end up with $13 < x$ as a solution, you can reverse the sides, but if you do so, you also need to reverse the inequality sign like this: $x > 13$. If you forget to do so, the arrow on your graph will end up pointing the wrong way.

Critical Point

A solid dot on a number line graph indicates that the given number should be included as a possible solution, whereas an open dot indicates that the given number cannot be a solution. For example, if you graph $x > 7$, you place an open dot at 7 because it's not a valid answer (7 is not greater than itself).

4. **It's not impolite to point.** Draw a dark arrow from your dot that points in the direction of the inequality symbol. If your final solution contained one of the less than symbols, draw the arrow to the left. If the solution contained a greater than symbol, the arrow (like the symbol) should point right.

Example 3: Graph the solution to $3(1 - 2x) < -5x + 6$.

Solution: First you need to solve the inequality for x. Remember, you want to get x on the left side of the inequality, so once you distribute the 3, separate the variable by adding $5x$ to and subtracting 3 from both sides.

$$3 - 6x < -5x + 6$$

$$3 - x < 6$$

$$-x < 3$$

Multiply (or divide) both sides by –1 in order to eliminate the negative sign in front of the x, and don't forget to reverse the inequality sign.

$$x > -3$$

To graph this solution, you'll draw an open dot (since the inequality symbol does not contain "or equal to") at –3 and sketch a dark arrow from that dot that extends to the right (since the inequality sign, like an arrowhead, points to the right), as shown in Figure 7.2.

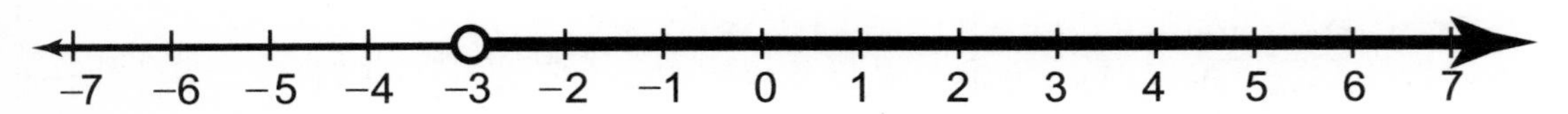

Figure 7.2

The graph of the inequality statement $x > -3$.

The graph makes it clear that every real number to the right of (but not including) –3 on the number line will make both the inequality $x > -3$ and therefore the original inequality, $3(1 - 2x) < -5x + 6$, true.

You've Got Problems

Problem 3: Graph the solution to $1 - x \geq 2x - 5$.

Compound Inequalities

You can define a range of values using a *single* inequality statement, called a *compound inequality*. Rather than say that, for a given equation, $x > 2$ and also $x \leq 9$, you could write $2 < x \leq 9$. This literally reads "2 is less than x, which is less than or equal to 9," but you might understand it better if I reword it like this: "x has a value between 2 and 9—it might even be 9, but it can't be 2."

Talk the Talk

A **compound inequality** is a single statement used to represent two inequalities at once, such as $a < x < b$. It describes a range (or *interval*) of numbers, whose lower boundary is a, and whose upper boundary is b. Whether or not those numbers are actually included in the interval is decided by the attached inequality sign. If the symbol is $\leq$, then the number is included; if it's $<$, then the boundary is not included.

In Figure 7.3, I illustrate the parts of a compound inequality. Notice that you should always write the lower boundary on the left and the upper boundary on the right. Furthermore, you should always use either the $<$ or $\leq$ symbol, or the statement may not make sense.

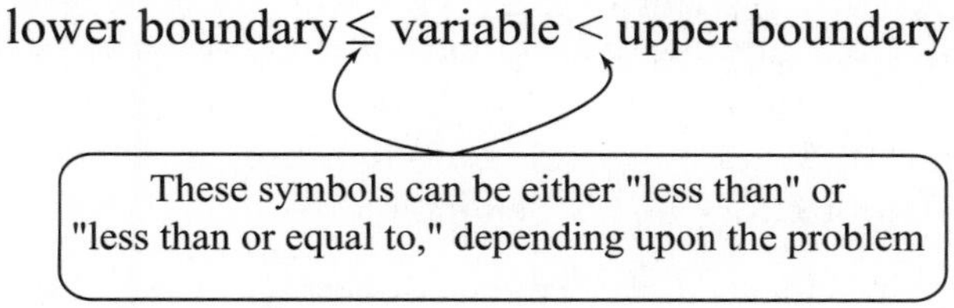

lower boundary ≤ variable < upper boundary

These symbols can be either "less than" or "less than or equal to," depending upon the problem

Figure 7.3

The anatomy of a compound inequality statement.

Solving Compound Inequalities

Notice that a compound inequality has three pieces to it, rather than the usual two sides of an equation or simple inequality. In order to solve a compound inequality, your job is to isolate the variable in the middle of the statement. Do this by adding, subtracting, multiplying, and dividing the same thing to *all three parts* of the inequality at the same time.

Kelley's Cautions

If you have to divide all three parts of a compound inequality by a negative number, you'll have to reverse *both* inequality signs to end up with this: $b > x > a$, where b is the upper boundary and a is the lower boundary. That's okay, but I prefer it rewritten as $a < x < b$, with less than signs and the lower boundary back on the left, where it belongs.

Example 4: Solve the inequality $-4 \leq 3x + 2 < 20$.

Solution: You want to isolate the x where the $3x + 2$ now is, so start by subtracting 2, not only from there, but from all three parts of the inequality.

$$-6 \leq 3x < 18$$

To eliminate the x's coefficient, divide *everything* by 3.

$$-2 \leq x < 6$$

This means any number between –2 and 6 (including –2 but excluding 6) will make the compound inequality true.

You've Got Problems

Problem 4: Solve the inequality $-1 < 2x + 5 < 13$.

Graphing Compound Inequalities

To graph a compound inequality, use dots to mark its endpoints on a number line. Just like in basic inequality graphs, the dots of compound inequalities correspond with the type of inequality symbol. Specifically, the symbol $\leq$ should be marked with a solid dot, and the symbol $<$ should be marked with an open dot. Since a compound inequality statement has two inequality symbols, use the one closest to each endpoint to help you decide what sort of dot to draw for that endpoint.

Once you've drawn the dots, draw a dark segment connecting them. This dark segment indicates that all the numbers between the endpoints are solutions to the inequality.

Example 5: Graph the compound inequality $-14 < x \leq -6$.

Solution: Since –14, the lower endpoint of this interval, has a $<$ symbol next to it, draw an open dot at –14 on the number line. The upper boundary, on the other hand, has a $\leq$ symbol next to it, so mark –6 with a solid dot. Now, connect the two dots with a dark line, as shown in Figure 7.4.

You've Got Problems

Problem 5: Graph the compound inequality $-2 \leq x < 5$.

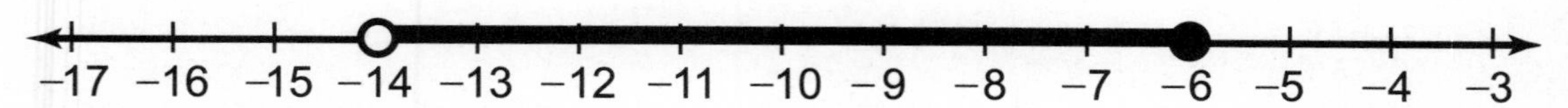

Figure 7.4

The graph of $-14 < x \leq -6$*. Note that this number line is not centered at 0; since all the key points on the graph are negative, it features those numbers instead.*

Inequalities with Absolute Values

What is it about absolute values? Just when you get used to doing something one way, along come these little teeny bars that demand things be done their way. They are just like that high-maintenance boyfriend or girlfriend, who wasn't content to let you live your life the way you always had. No, suddenly it was "Why can't we eat somewhere nice for a change?" or "Is there a federal law against you brushing your teeth more than once a month?" Not that these adjustments were difficult to make, it's just that sometimes having to change little things once you get into a routine can be tricky.

Back in Chapter 5, you had to split absolute value equations into two distinct, nonabsolute value equations in order to reach a solution. Similarly, you have to break absolute value inequalities into two distinct, nonabsolute value inequalities to reach a solution. However, just to make things a little worse (if that were possible), the process is different for inequalities with less than symbols and those containing greater than symbols.

Inequalities Involving "Less Than"

If you're asked to solve an inequality problem containing absolute values, and the inequality symbol is either $<$ or $\leq$, here are the steps you should follow to reach a solution:

1. **Isolate the absolute value portion on the left side of the inequality.** When you do, the problem should look something like this: $|x+a| < b$, where a and b are real numbers. (If a is negative, the problem will look like $|x-a| < b$, and that's okay, too.)
2. **Create a compound inequality.** Rewrite the statement $|x+a| < b$ as $-b < x + a < b$. In other words, drop the absolute value symbols, place a matching inequality symbol to the left of the statement, and then write the opposite of the constant to the left of that.
3. **Solve the compound inequality.** Use the procedures I showed you in the previous section to solve and/or graph the compound inequality.

Example 6: Solve the inequality $|2x-1|+3 \le 6$ and graph the solution.

Solution: To isolate the absolute value quantity, subtract 3 from both sides.

$$|2x-1| \le 3$$

Drop the absolute value bars, and write the opposite of 3 to the left of the expression. Between the newly added –3 and now "bar-free" expression, place a ≤ symbol, to match the one already there.

$$-3 \le 2x - 1 \le 3$$

Solve the compound inequality by adding 1 to each part and then dividing everything by 2.

$$-2 \le 2x \le 4$$

$$-1 \le x \le 2$$

Graph the solution by placing solid dots at –1 and 2 and connecting them, as shown in Figure 7.5.

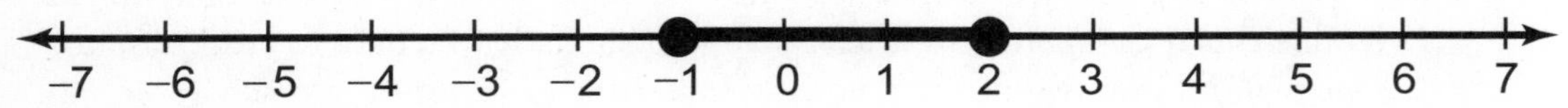

Figure 7.5

The graph of $|2x-1|+3 \le 6$ *is the same as the graph of the compound inequality* $-1 \le x \le 2$.

If this problem had contained < symbols instead of ≤, the procedure would have been exactly the same, just with those less than signs throughout; of course, the graph would have contained open dots, since they coordinate so nicely with those less than signs.

You've Got Problems

$$4|x-5| < 8$$

Inequalities Involving "Greater Than"

Absolute value inequalities containing the symbols > or ≥ are solved much like their "less than" inequality sister problems. For one thing, you must start by isolating the absolute value quantity first, and then you must rewrite the expression without absolute value bars. However, this time you won't end up with a compound inequality, because you must rewrite the expression differently.

Once you've isolated the absolute value quantity, rewrite the statement as two separate inequalities, one that looks just like the original (just without absolute value bars), and the other with the inequality sign reversed and the opposite of the constant. Basically, the expression $|x+a|>b$ should be rewritten as the two expressions

$$x + a > b \text{ or } x + a < -b$$

Note the word "or" between the expressions. That doesn't mean that you only have to write one *or* the other (they both need to be in your solution); it means that if you plug an x into *either* expression and it works for *just one* of them, then that x value is a solution to the original inequality.

Example 7: Solve the inequality $|2x+5|-4>-1$ and graph the solution.

Solution: Start by isolating the absolute value quantity (add 4 to both sides).

$$|2x+5|>3$$

Split this into two separate inequality statements. Create the first simply by removing the bars; the second requires you to reverse the inequality sign and take the opposite of that lone constant 3. (It's good form to write the word "or" between the statements.)

$$2x + 5 > 3 \text{ or } 2x + 5 < -3$$

Now solve each basic inequality separately.

$$2x > -2 \text{ or } 2x < -8$$

$$x > -1 \text{ or } x < -4$$

That weird-looking two-headed monster is the answer. It says that any number greater than –1 *or* less than –4 will make the original inequality true. Since the solution is made up of two basic inequality statements, the graph of the solution will consist of the graphs of those statements put together. Just graph both of the inequalities on the same number line, as demonstrated in Figure 7.6.

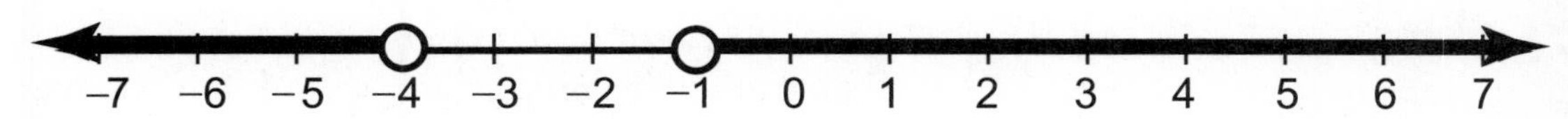

Figure 7.6

The solution graph to $|2x+5|-4>-1$ *is the same as the graphs of* $x>-1$ *and* $x<-4$ *put together.*

You've Got Problems

Problem 7: Solve the inequality $|x-4|\geq 2$ and graph the solution.

Graphing Linear Inequalities

Wouldn't it be great if you suddenly discovered that you had a skill you didn't know about? What if one day, while you're out jogging, you found out that by twisting your right leg just a little, you could suddenly speak Portuguese fluently? (Assuming of course that you couldn't speak Portuguese before.) That'd really be something to write home about! (Perhaps even in Portuguese.) Well, you're about to find out that (with just a slight twist of your right leg) you can graph linear inequalities.

The difference between linear inequalities and the inequalities you've dealt with so far in this chapter is that the linear ones have two variables, usually x and y. The great thing about these linear inequality graphs is that they are based on the graphs of linear equations, which you already know how to create. However, if you compare the graphs of linear equations with the graphs of linear inequalities, these are the major differences you'll see in the inequality graphs:

- **The graph is not always a solid line.** If the inequality symbol in the statement is either $<$ or $>$, then the line in the inequality graph will be dotted, not solid, to indicate that the points along the line are not solutions to the inequality. However, if the inequality symbol is $\leq$ or $\geq$, the line will be solid, just like you're used to.
- **The solution is not just the line, it's an entire region of the graph.** The lines (whether solid or dotted) split the coordinate plane into two parts, sort of the way a fence divides up a piece of property. All of the points on one side of the line will make the inequality true, whereas all the points on the other side will not. You indicate the region of solutions by lightly shading it on your graph.

Critical Point

The dotted line in a linear inequality is the two-dimensional equivalent of the open circle in a basic inequality. Both mean "don't include this value (or coordinate pair) as a possible solution."

Therefore, to graph a linear inequality, first treat it as though it were a line, and graph that line. (Make sure to check whether the line should be solid or dotted, based on the inequality symbol.) Then choose a coordinate, called a *test point*, from the coordinate plane and plug it into the original inequality. If it makes the inequality true, then so will all the other points on the same side of the line you drew, so you should shade in that region. If it doesn't work, then the region on the other side of the line is your solution.

Example 8: Graph the inequality $2x - 3y < 12$.

Solution: For just a moment, pretend that the inequality symbol is an equal sign, and graph the resulting linear equation. (Since the inequality does not indicate "or equal to," make the line dotted, as shown in Figure 7.7.) The easiest way to graph the line is to calculate its intercepts using the method of Chapter 5.

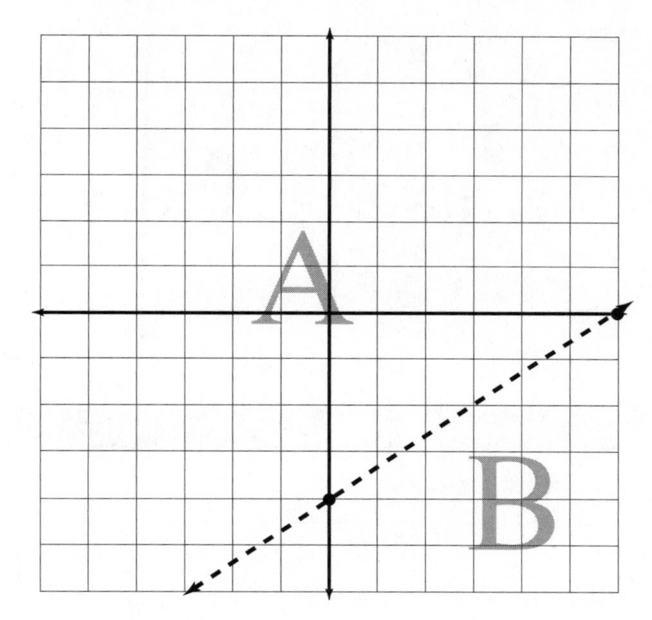

Figure 7.7

This is not yet the final graph, but notice how the line splits the coordinate plane into two regions, labeled A *and* B *for clarity.*

Now you should choose a test point from one of the two regions. I usually choose the origin, because what could be easier than plugging in 0 for both x and y? Whatever point you choose, plug in its x-coordinate for x in the inequality and its y-coordinate for y. Here's how it works for the origin:

$$2(0) - 3(0) < 12$$
$$(0) < 12$$

Because this statement is true (0 is, indeed, less than 12), the origin is a solution point, and so is every other point in its region of the plane (region A, in Figure 7.7). So, shade in that portion of the graph to get your final graph (see Figure 7.8).

Figure 7.8

The graph of $2x - 3y < 12$ in all its splendor. Looks shady to me ….

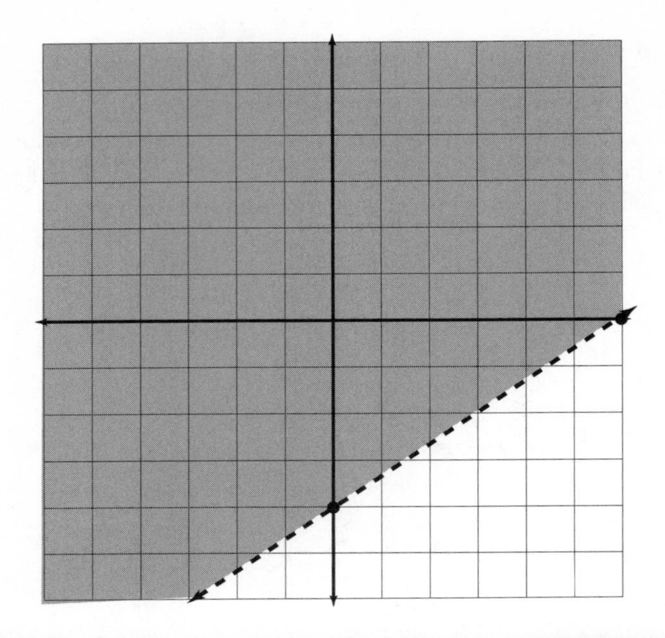

You've Got Problems

Problem 8: Graph the inequality $y \geq -2x + 3$.

The Least You Need to Know

- If you multiply or divide an inequality by a negative number, you must reverse the inequality sign.
- It is important to demonstrate (via open dots or dotted lines) when points or lines are not included in a graph, as a result of an inequality symbol of < or >.
- Compound inequalities express two inequalities in one statement.
- Linear inequality graphs contain shaded solution regions.

Part 3

Systems of Equations and Matrix Algebra

You're about to solve bunches of equations, not one at a time, but simultaneously! (Forget mathematician! That sounds like the work of a mathe*magician.*) In this part, you'll use lots of different methods to solve groups of equations, including a foray into the land of the matrix. (Unfortunately, that's not as cool as it sounds.)

Chapter 8

Systems of Linear Equations and Inequalities

In This Chapter

- Defining systems of equations
- Solving systems using different techniques
- Classifying systems with infinite or no solutions
- Graphing solutions to systems of inequalities

What could be more fun than one linear equation, you ask? (I know you didn't actually ask that and never would of your own volition, but play along.) Why, of course, the answer would be *two* linear equations! By now you know that linear equations have an infinite number of solution pairs, each of which makes the linear equation true.

Now, however, instead of just one measly equation, I'll give you two equations at once (called a *system of equations*). Can you believe it? Two equations for the price of one, all for seven easy payments of $19.95—plus I'll throw

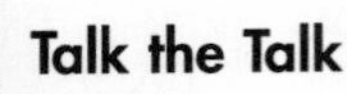

Talk the Talk

A **system of equations** is a collection of two or more equations. They're usually written with a left brace to indicate that they all belong to the same system, like so:

$$\begin{cases} x+y=7 \\ 2x-1=4 \end{cases}$$

in this contraption that lets you make your own beef jerky at home! Your job will be to tell me what coordinate pair or coordinate pairs make *both* of the equations true.

Unlike some algebra problem types, there are a few different techniques you can use to reach that answer, so you'll actually have a choice between them when it comes time to solve these systems. However, some are easier to solve using one technique versus another, so make sure you are able to use them all effectively.

Solving a System by Graphing

Remember, a graph is a visual representation of all the solutions to a linear equation. Therefore, if you draw the graphs of both linear equations in a system on a single coordinate plane, any point at which those graphs overlap will represent a solution the equations have in common. Since the graphs of linear equations are just lines, and usually two lines intersect at a single point, most of the systems of linear equations you'll deal with will have a single answer.

Critical Point

Notice that I said *usually* two lines intersect at a point; sometimes it doesn't work like that. What if two lines are parallel (so they never intersect) or two lines completely overlap one another (so they intersect at every single point)? Well, you'll have either no solutions or more than you can count. I'll explain how to handle both later on in the chapter.

In other words, all you have to do to solve a system of equations is find the intersection point of two lines (and you already know how to graph lines). The hard part, though, is making your graphs precise. Since your answer depends on how exact your graphs are, you need to be extra careful when drawing them. I recommend you use graph paper and a ruler, to make sure your measurements and lines are perfect.

Example 1: Solve the system of equations by graphing.

$$\begin{cases} x-2y=8 \\ -2x-y=-6 \end{cases}$$

Critical Point

You can write the solution to the system in Example 1 as either (4,–2) or $x = 4$, $y = 2$.

Solution: I suggest graphing using intercepts. The first equation of the system will have intercepts (8,0) and (0,–4); the second equation will have intercepts (3,0) and (0,6). As you can see in Figure 8.1, the graphs overlap at the point (4,–2).

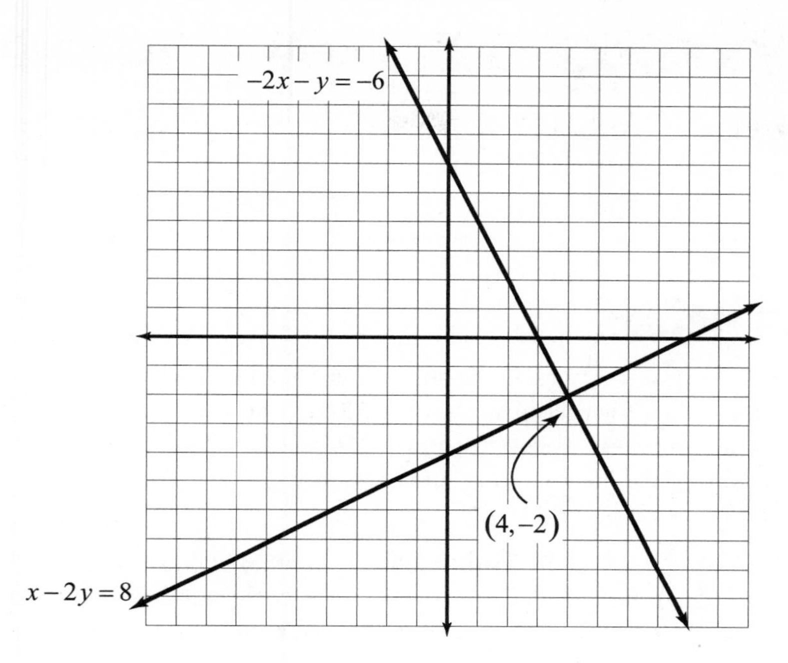

Figure 8.1

The solution to the system of equations is the point at which the graphs of both equations in the system intersect.

Actually, it would be more accurate to say that it *looks like* the graphs in Figure 8.1 overlap at (4,–2); after all, you'll have to produce these graphs by hand, so they won't possess the same precision as my computer-drawn graphs. To make sure the solution to the system actually is that point, plug in 4 for x and –2 for y into both of the equations in the system; if you get true statements once you simplify, you can rest assured that you got the answer correct.

$$\begin{aligned} x - 2y &= 8 \\ 4 - 2(-2) &= 8 \\ 4 + 4 &= 8 \\ 8 &= 8 \end{aligned} \qquad \begin{aligned} -2x - y &= -6 \\ -2(4) - (-2) &= -6 \\ -8 + 2 &= -6 \\ -6 &= -6 \end{aligned}$$

Lucky for us, the coordinates were both integers, which are easily identified on the coordinate plane. On the other hand, if the solution to the system had been something much uglier, like $\left(4\frac{1}{19},-2\frac{3}{8}\right)$, there would have been no way to get that answer simply by looking at the intersection point on a hand-drawn graph. Therefore, this method of solving systems of equations is not used very often; more sophisticated methods, which you'll learn in the next couple of sections, are necessary.

You've Got Problems

Problem 1: Solve the system by graphing.

$$\begin{cases} 2x+y=3 \\ x-3y=9 \end{cases}$$

The Substitution Method

If one of the equations in a system can easily be solved for one of its variables, then that system is a prime candidate for the substitution method. The great thing about this alternative to the graphing method you just learned is that you don't need any new skills—you just need to know how to solve an equation for a variable, and you learned that in Chapter 4.

By the way, the substitution method can be used to solve any system of equations, and if the system has a solution, this technique will *always* get you the correct answer. However, the more complicated the equations in the system, the more arithmetic goes into the process, and the easier it is to make a mistake along the way.

The best time to use the substitution method is when you notice that one of the variables in the system has a coefficient of 1 or –1. (In other words, there is no numeric coefficient written next to the variable at all.) That "absent" coefficient makes it easy to solve for its corresponding variable, and the substitution method is easier than falling down the stairs (which I know for a fact is a very easy thing to do, considering the amazing number of times I have managed to do it).

CAUTION

Kelley's Cautions

Once you solve one of the equations in the system for a variable, make sure to plug the result into the *other* equation. If, instead, you plug the result back into the original form of the equation you solved, all the terms will cancel out, and you'll eventually get 0 = 0, a true but unhelpful statement.

Example 2: Use the substitution method to solve the system.

$$\begin{cases} 7x+4y=6 \\ 2x-y=-9 \end{cases}$$

Solution: Since the coefficient of y in the second equation is –1, it makes solving for y pretty simple. Just subtract $2x$ from both sides in that equation and multiply everything by –1.

$$-y = -2x - 9$$

$$-1(-y) = -1(-2x - 9)$$

$$y = 2x + 9$$

How'd You Do That?

The substitution method works by eliminating one of the variables in an equation. By solving the second equation for y in Example 2, you are actually rewriting y as a statement containing x's. So, when you replace the y with x's in the other equation, the result is an equation with all x's in it, which will have only one answer; this is much better than the original linear equation, which (like all linear equations) has an infinite number of coordinate pair solutions.

Now that you have solved for y, plug that *entire quantity* $(2x + 9)$ in place of y in the other equation.

$$7x + 4y = 6$$

$$7x + 4\,(2x + 9) = 6$$

Distribute the 4 and solve for x.

$$\begin{aligned} 7x + 8x + 36 &= 6 \\ 15x + 36 &= 6 \\ 15x &= -30 \\ x &= -2 \end{aligned}$$

You're halfway to the correct answer—remember the solution to a system is the coordinate pair (x,y) where the graphs overlap. To find the corresponding y-value, plug the x-value you just got into the equation you originally solved for y.

$$y = 2x + 9$$

$$y = 2(-2) + 9$$

$$y = -4 + 9$$

$$y = -5$$

That's all there is to it. The solution to the system is (–2,5), which you could also write as $x = -2$, $y = 5$.

You've Got Problems

Problem 2: Use the substitution method to solve the system.

$$\begin{cases} x - 4y = 11 \\ 3x + 7y = -5 \end{cases}$$

The Elimination Method

As I intimated in the last "How'd You Do That?" note, the substitution method is basically a sneaky way to eliminate a variable. It's the boxing equivalent of a sucker punch; before the system knew what hit it, wham! It's over! One of the variables is lying unconscious in the boxing ring, and you can handle the variable that's left with no trouble. (After all, two against one isn't fair, is it?)

If the substitution method is a sucker punch, then the other alternative you have for solving systems of equations, called the *elimination method*, is the metaphorical equivalent of a full-blown wrestling match. There's no finesse or devious tactical plan at work here. You simply stride in, manhandle the equations with brute force, and show them you're in charge. Like the substitution method, elimination will work for any system that has a valid solution (and will even shed light on the answer if there isn't one, as you'll learn later in the chapter).

As the name suggests, the whole purpose of the elimination method is to eliminate a variable, by multiplying one or both of the equations in the system by a real number. Multiplying by a real number, as long as you do it to both sides of the equation, won't change the equation's solution, so you may be wondering how that helps. Well, you don't just multiply willy-nilly; there's a trick to it. You multiply so that if you were to line up the like terms in both equations and add those equations together, one of the variables would vanish.

Once that variable's gone, the wrestling match is basically over. All that remains, before you can pin your opponent, is to solve the equation and then evaluate the missing variable.

Critical Point

If you're having trouble figuring out what to multiply each equation by to eliminate a variable, use this little trick. First, write the equations in standard form (if they're not already). You'll end up with something like this:

$$\begin{cases} ax + by = c \\ dx + ey = f \end{cases}$$

(Of course, a, b, c, d, e, and f will actually be real number coefficients and constants.) To eliminate x, multiply the top equation by d and the bottom equation by $-a$.

For example, instead of performing the multiplication I used in Example 3, you could have multiplied the top equation by 6 and the bottom equation by -2, and you would have gotten the same final answer.

Example 3: Solve the system using the elimination method.

$$\begin{cases} 2x - y = 13 \\ 6x + 4y = 4 \end{cases}$$

Solution: Notice that if you multiply the entire top equation by –3, you'll get $-6x + 3y = -39$. Why would you want to do that? The *x* coefficient would be –6, the opposite of the *x* coefficient in the other equation, 6. If you were to add the new version of the top equation to the bottom equation, the *x*-terms will disappear.

$$\begin{array}{rrrcr} & -6x & +3y & = & -39 \\ + & 6x & +4y & = & 4 \\ \hline & & 7y & = & -35 \end{array}$$

The resulting equation is very simple to solve; just divide both sides by 7, and you get $y = -5$. Now it's time to figure out what the *x*-value is that, when paired with this *y*-value, completes the ordered pair of the solution. Just plug the value $y = -5$ into *either* of the linear equations from the problem; you'll get the correct answer no matter which one you pick.

I'll substitute $y = -5$ into the first equation in the original system, $2x - y = 13$.

$$2x - (-5) = 13$$

$$2x + 5 = 13$$

$$2x = 8$$

$$x = 4$$

The solution to this system is the coordinate pair (4,–5). (Make sure always to write the *x*-value before the *y*-value in an ordered pair.)

CAUTION

Kelley's Cautions

When multiplying an equation by a number, don't forget the constant. In other words, if asked to multiply the equation $2x - y = 13$ by –3, don't answer $-6x + 3y = 13$; students often forget to multiply the 13 by –3, since it's all by itself on the other side of the equal sign.

You've Got Problems

Problem 3: Solve the system using the elimination method.

$$\begin{cases} 2x - y = -11 \\ 5x + 2y = 4 \end{cases}$$

Systems That Are Out of Whack

Occasionally, as you attempt to solve a system of equations, something very bizarre will happen—when you open your algebra book, it will suck you in and transport you to the magical world of Narnia. Actually, that's not true. The weird happenstance is this: All of the variables disappear!

When all the variables vanish, it means one of two things has happened:

- **The system has no solutions.** If the statement you end up with is false, such as $0 = 7$, then the linear equations in the system have no common solutions, and the system is said to be *inconsistent.* This happens when the graphs of the lines are parallel to one another, so they never intersect at a solution point.
- **The system has infinitely many solutions.** If the statement that results is true, such as $5 = 5$, then the linear equations in the system have graphs that overlap, so they share *every* point in common, not just one. Such systems are called *dependent.*

Talk the Talk

A system of linear equations that has no solutions is called **inconsistent,** whereas a system that has infinitely many solutions is said to be **dependent.**

Example 4: Solve the system of equations.

$$\begin{cases} x - 2y = 1 \\ -3x + 6y = 3 \end{cases}$$

Solution: To solve this system via the elimination method, multiply the first equation by 3; the x-terms will cancel when the equations are added.

$$\begin{array}{rrcr} 3x & -6y & = & 3 \\ -3x & +6y & = & 3 \\ \hline & 0 & = & 3 \end{array}$$

Hey wait a minute! The y-terms vanished, too! Since all the variables are gone, and the statement $0 = 3$ is false, you classify the system as inconsistent; there are no solutions.

You would have gotten the same answer had you tried to solve this system using the substitution method. If you solve the first equation for x, you get $x = 2y + 1$. When you plug that into the second equation, you'll end up with a different but still false, variable-free statement.

$$\begin{aligned} -3(2y + 1) + 6y &= 3 \\ -6y - 3 + 6y &= 3 \\ -3 &= 3 \end{aligned}$$

You've Got Problems

Problem 4: Solve the system of equations.

$$\begin{cases} 4x + 3y = -2 \\ -8x - 6y = 4 \end{cases}$$

Systems of Inequalities

In Chapter 7, you learned that the graph of a linear inequality is a shaded region of the coordinate plane, bounded by a line which is either solid or dotted, depending upon the inequality symbol involved. The graph of a system of inequalities is not much different than that.

Here's how to graph the solution to a system of inequalities:

1. **Graph each inequality in the system on the same coordinate plane.** Make sure to shade lightly; if there are numerous inequalities in the system, the graph can get messy fast.
2. **The solution is the overlapping shaded area.** The region on the graph containing shading *from every inequality in the system* is your final solution.

The common shaded region is the solution to the system because it contains the points whose coordinates make *all* of the inequalities in the system true.

Example 5: Graph the solution to the system of inequalities.

$$\begin{cases} x > -3 \\ x - 2y \leq 4 \end{cases}$$

Solution: Graph both inequalities on the same coordinate plane. (In case you're wondering how to graph $x > -3$ since it has only one variable, treat it like the equation $x = -3$, the vertical line three units left of the y-axis.)

You can then choose a test point and proceed as you did in Chapter 7. Don't forget to make sure that you're using solid and dotted lines for the graphs appropriately. You should end up with the graph in Figure 8.2.

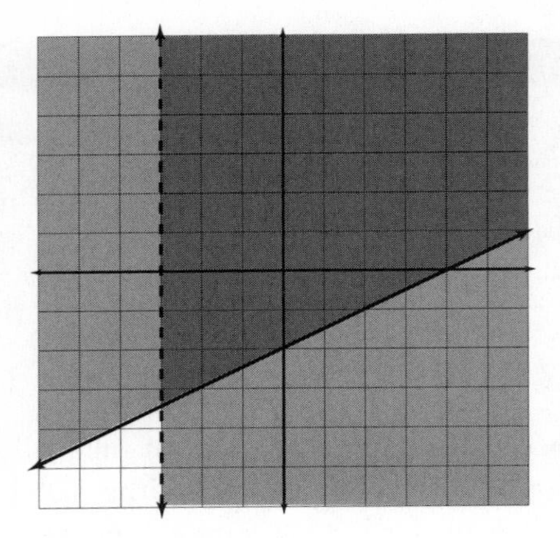

Figure 8.2

Use test points to determine that you should shade to the right of the vertical line $x > -3$ *and above the line* $x - 2y \leq 4$.

The darker shading in Figure 8.2 is the solution region for the system of inequalities, since it is the only region where the shading for both inequalities overlaps. Some teachers will allow you to leave the graph as is, accepting the fact that the darker shading represents the final answer, whereas some would rather you graph *only* the solution, as pictured in Figure 8.3.

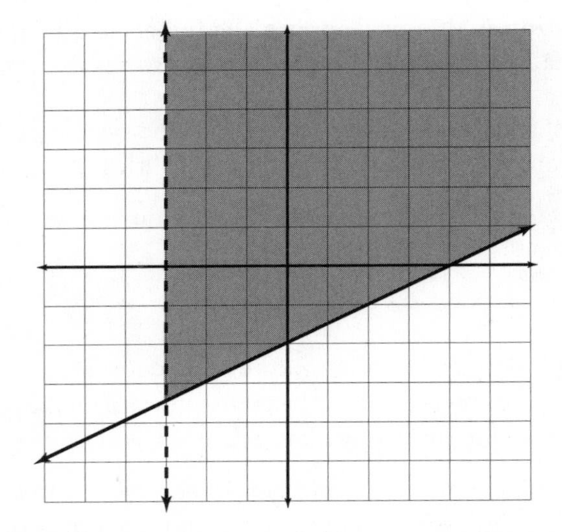

Figure 8.3

It's easier to identify the solution to the system of inequalities in this graph than in Figure 8.2.

You've Got Problems

Problem 5: Graph the solution to the system of inequalities.

$$\begin{cases} y \leq -\dfrac{2}{3}x + 1 \\ y > 4x - 5 \end{cases}$$

The Least You Need to Know

- The solution to a system of equations makes all of the equations in the system true.
- A linear system of equations will have either one solution, no solutions (in inconsistent systems), or infinitely many solutions (in dependent systems).
- The elimination and substitution methods allow you to solve systems of equations by forcing them to contain only one variable.
- Systems of linear inequalities have solutions which are represented as shaded regions on a coordinate plane.

Chapter 9

The Basics of the Matrix

In This Chapter

- Defining a matrix and its component parts
- Adding, subtracting, and multiplying matrices
- Calculating determinants of matrices
- Solving systems of equations using determinants

When most people hear the word "matrix" nowadays, they think about people in dark suits who can jump from building to building, avoid bullets in slow motion while contorting at impossible angles, and that guy who was in the movie *Bill and Ted's Excellent Adventure.*

There are a few differences between a mathematical matrix and a matrix as described by the science fiction movie. For one thing, a math matrix contains almost no violence (none, in fact, unless you punch someone while trying to figure out your homework problems). Second of all, there are no attractive people skulking about in a math matrix—no beautiful women or strapping hunks wearing shiny vinyl and fighting for the life of mankind. Just a bunch of numbers written in rows and columns. In fact, the most attractive person in a math matrix is probably your sixth-grade math teacher, and he wore nothing that wasn't fashioned out of brown polyester.

While the realm of the mathematical matrix may not be exciting, death-defying, or packed with stunts that cost millions of dollars, they are still pretty neat. They're completely different from both the stuff you have been working on in preceding chapters and the stuff you'll see in the chapters that follow. As alien as they are, they have surprising uses; in fact, you'll even use them to solve systems of equations (in case the three techniques you learned in the last chapter didn't satisfy you).

What Is a Matrix?

Simply put, a *matrix* is a group of objects called *elements* (which are usually numbers) arranged in rows and columns and surrounded with brackets. The *order* of a matrix describes the number of rows and columns that matrix contains. Consider matrix A as defined below.

$$A = \begin{bmatrix} 2 & -1 & 3 & 0 & 9 \\ -8 & -6 & 4 & 7 & 13 \\ -20 & -3 & 8 & 11 & 1 \end{bmatrix}$$

Matrix A has order 3×5 (read "3 by 5"), since it contains three rows and five columns. (Rows are horizontal lines of numbers and columns are vertical lines of numbers; when writing the order of a matrix, you always list the number of rows before the number of columns.) If a matrix has an equal number of rows and columns, it is said to be a *square matrix*.

If you are discussing individual elements of a matrix, you use the notation a_{rc}, where r is the element's row, and c is its column. For example, in matrix A above, $a_{24} = 7$ since the number 7 appears in the second row and fourth column of the matrix. Notice that the lowercase variable used to represent the elements matches the uppercase variable of the matrix's name (a and A).

Talk the Talk

A **matrix** is a rectangular collection of numbers, called **elements,** arranged in rows and columns and surrounded by brackets. A matrix containing m rows and n columns is said to have **order** $m \times n$; if the matrix has the same number of rows and columns, it is described as **square.**

Critical Point

You can remember that *columns* are *vertical* because they both contain the letter *c*, or just look at the Lincoln Memorial on the back of a penny—it features numerous *vertical columns* in its architecture.

Matrix Operations

Now that you know what a matrix is, it's time to start working with them. Although there are entire college math courses dedicated to matrices and how dang useful they can be, in algebra, you'll only be expected to perform a few matrix operations (like a tracheotomy and a tonsillectomy, for starters).

Multiplying by a Scalar

The simplest matrix operation requires you to multiply all of the numbers in the matrix by a separate real number, called a *scalar*. It's sort of like the distributive property on a grand scale; every number in the matrix gets multiplied by the scalar, and you just write the product in the same order as the original matrix.

Talk the Talk

A **scalar** is a real number. You use the term "scalar" to designate a real number not contained within the matrix, as opposed to an "element," which is a number holding a spot inside the matrix.

Example 1: If $A = \begin{bmatrix} -1 & 3 & 6 \\ -4 & 0 & -2 \end{bmatrix}$, evaluate $-2A$.

Solution: This problem asks you to multiply all of the elements in matrix A by the scalar value –2. It's a very simple task—just be careful with the signs as you multiply.

$$-2A = \begin{bmatrix} (-2)(-1) & (-2)(3) & (-2)(6) \\ (-2)(-4) & (-2)(0) & (-2)(-2) \end{bmatrix}$$

$$-2A = \begin{bmatrix} 2 & -6 & -12 \\ 8 & 0 & 4 \end{bmatrix}$$

You've Got Problems

Problem 1: If $B = \begin{bmatrix} 1 & 4 & 9 \\ -2 & 7 & -3 \\ 11 & -5 & 6 \end{bmatrix}$, evaluate $4B$.

Adding and Subtracting Matrices

In order to add two matrices together, all you do is find the sum of their corresponding elements. In other words, if you're asked to add matrices A and B together to get matrix C, go element by element. Add the top left element in matrix A to the top left element of matrix B, and write the result in the top left element spot of matrix C. Repeat this process for every position in the matrices, and you're done.

Kelley's Cautions

You can only add or subtract matrices A and B if they have the same order. If one has more rows or columns than the other, neither the sum nor the difference of the matrices is defined.

Subtracting matrices is no different; you just use the word "subtraction" if you happen to be multiplying one of the matrices by a negative scalar value. If you're a little unsure about what I mean, check out Example 2(b).

Example 2: If $A = \begin{bmatrix} 1 & 5 & -9 \\ 6 & -4 & 13 \\ 2 & -2 & 3 \end{bmatrix}$ and $B = \begin{bmatrix} 10 & 8 & -5 \\ 4 & 3 & -1 \\ 0 & 2 & 6 \end{bmatrix}$, evaluate the following expressions.

(a) $A + B$

Solution: Both matrices have nine elements; just add each element in A to the element in B in the same corresponding spot and write the result in the same corresponding position in the answer matrix.

$$A + B = \begin{bmatrix} 1+10 & 5+8 & -9+(-5) \\ 6+4 & -4+3 & 13+(-1) \\ 2+0 & -2+2 & 3+6 \end{bmatrix}$$

$$A + B = \begin{bmatrix} 11 & 13 & -14 \\ 10 & -1 & 12 \\ 2 & 0 & 9 \end{bmatrix}$$

(b) $2A - B$

Solution: In this expression, the elements of A are multiplied by a scalar value of 2, and the elements of B are multiplied by a scalar value of –1. Start by performing the scalar multiplication.

$$2A = \begin{bmatrix} 2 & 10 & -18 \\ 12 & -8 & 26 \\ 4 & -4 & 6 \end{bmatrix} \qquad -B = \begin{bmatrix} -10 & -8 & 5 \\ -4 & -3 & 1 \\ 0 & -2 & -6 \end{bmatrix}$$

Don't think to yourself, "I need to subtract matrices to get $2A - B$." Instead, think, "To get $2A - B$, I should *add* the matrices $2A$ and $-B$."

$$2A - B = \begin{bmatrix} 2 & 10 & -18 \\ 12 & -8 & 26 \\ 4 & -4 & 6 \end{bmatrix} + \begin{bmatrix} -10 & -8 & 5 \\ -4 & -3 & 1 \\ 0 & -2 & -6 \end{bmatrix}$$

$$2A - B = \begin{bmatrix} 2+(-10) & 10+(-8) & -18+5 \\ 12+(-4) & -8+(-3) & 26+1 \\ 4+0 & -4+(-2) & 6+(-6) \end{bmatrix}$$

$$2A - B = \begin{bmatrix} -8 & 2 & -13 \\ 8 & -11 & 27 \\ 4 & -6 & 0 \end{bmatrix}$$

You've Got Problems

Problem 2: If $A = \begin{bmatrix} -1 & \frac{1}{2} \\ 5 & 3 \end{bmatrix}$ and $B = \begin{bmatrix} 1 & -4 \\ -7 & 6 \end{bmatrix}$, evaluate $4A - 3B$.

Multiplying Matrices

The trickiest thing you'll need to do with matrices is to calculate their products. Unfortunately, matrix multiplication does not work like matrix addition and subtraction. In other words, you can't just multiply corresponding pieces of matrices to get the solution.

When Can You Multiply Matrices?

One of the biggest differences between matrix multiplication and matrix addition is that matrix multiplication does not require that the two matrices involved have the same order. Instead, multiplication has its own requirement: The number of *columns* in the first matrix must equal the number of *rows* in the second matrix. In other words, if matrix A has order $m \times n$ and matrix B has order $p \times q$, in order for the product $A \cdot B$ to exist, n and p must be equal.

Here's another little nugget: If the product of two matrices exists, it will have the same number of rows as the first matrix and the same number of columns as the second. In other words, if matrix A has order $m \times n$ and matrix B has order $n \times p$, $A \cdot B$ will have order $m \times p$.

Kelley's Cautions

Example 3 demonstrates a very important point—matrix multiplication is not commutative. Just because the matrix product $A \cdot B$ exists, that does not guarantee that $B \cdot A$ does.

Example 3: If matrix C has order 4×5 and matrix D has order 5×9, does the product $C \cdot D$ exist? If so, describe the order of the product. Does the product $D \cdot C$ exist? If so, describe its order.

Solution: Remember, for a matrix product to exist, the number of *columns* of the left matrix must equal the number of *rows* of the right matrix. ("Left" and "right" refers to the order in which the product is written.) This is true for $C \cdot D$, so the product exists. Furthermore, $C \cdot D$ will have order 4×9, the same number of rows as C and the same number of columns as D.

On the other hand, $D \cdot C$ does not exist, because the number of columns in D (9) does not equal the number of rows in C (4).

You've Got Problems

Problem 3: If A is a matrix of order 3×4 and B is a matrix of order 3×3, which product exists, $A \cdot B$ or $B \cdot A$? What is the order of the only possible product matrix?

Use Your Fingers to Calculate Matrix Products

If matrix A and B meet the row/column requirement and their product exists (I'll call the product matrix P), quite a bit of work goes into calculating each element in P. The best way to learn this process is via an example. All you'll need is a little patience, a couple of fingers, and the Ring of Power forged in the fiery bowels of Mordor (but only if you are Frodo Baggins; otherwise, the last requirement does not apply to you).

Example 4: Calculate the product.

$$\begin{bmatrix} 2 & -2 \\ 1 & 7 \end{bmatrix} \cdot \begin{bmatrix} -4 & 0 & -9 \\ 5 & 3 & -6 \end{bmatrix}$$

Solution: Since the first matrix is 2×2 and the second is 2×3, the product exists, and its dimensions will be 2×3. Start by writing a very generic 2×3 matrix.

$$\begin{bmatrix} p_{11} & p_{12} & p_{13} \\ p_{21} & p_{22} & p_{23} \end{bmatrix}$$

(Each term is written in the form I mentioned before, p_{rc}, where r indicates the element's row and c indicates its column.)

Here's the key to multiplying matrices: Each term, p_{rc}, is the result of multiplying the numbers in the rth row of the first matrix by the corresponding numbers in the cth column of the second and adding up all the results. Trust me, I know that sounds really weird and complicated, but it's not.

The two little numbers after each p in the generic product matrix tell you where to put your fingers when you calculate the answer. The *left* little number tells you to put your *left* index finger on the *left*most number in that numbered row of the *left* matrix. The *right* little number tells you to put your *right* index finger on the *top* of that numbered column in the *right* matrix.

You'll need to do this for every p term in the product matrix, but as an example, I'll calculate the p_{13} term. Since the left little number is 1, put your left index finger on the leftmost term in the *first* row of the left-hand matrix (the element 2 in this example). Since the right little number is 3, put your right index finger on the top number in the *third* column of the second matrix (the element –9 in this example). The correct finger positions are shown in Figure 9.1.

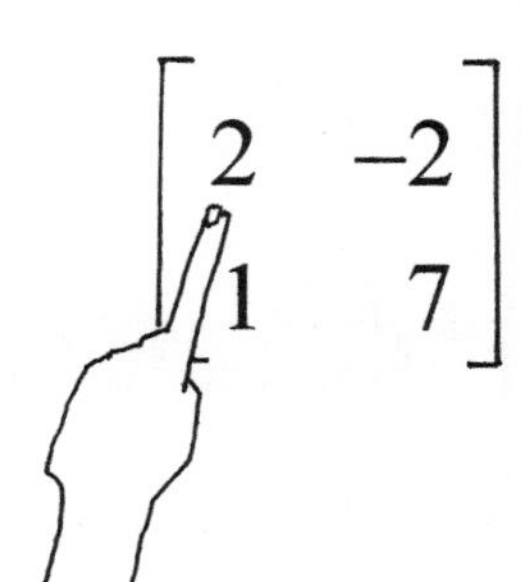

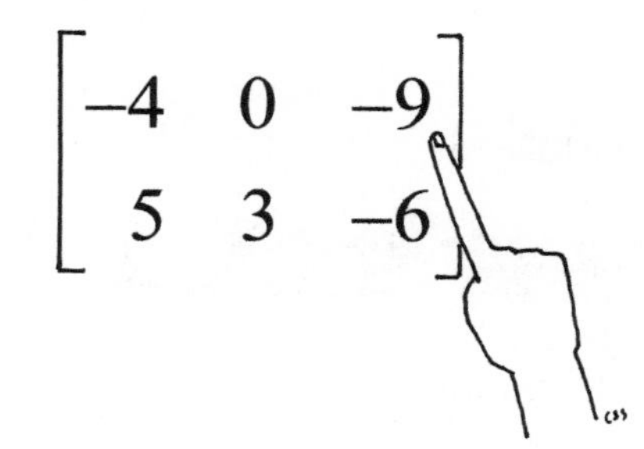

Figure 9.1

To calculate p_{13}, *your left and right index fingers should hover over the matrices like this.*

To actually calculate p_{13}, multiply the numbers you're pointing at, $2(-9) = -18$. Now move your left finger to the *right* along the row to the next number, –2, and move your right finger *down* along the column to the next number, –6, and multiply those numbers together: $(-2)(-6) = 12$. Do this until you simultaneously reach the end of the row in the left matrix and the column in the right matrix, and then add together all the products you got along the way:

Critical Point

Remember that in order for a matrix product to exist, the number of columns of the first and the rows of the second had to match? This guarantees that you'll reach the end of a row and the bottom of a column simultaneously when you multiply.

$$p_{13} = -18 + 12$$

$$p_{13} = -6$$

So, that's one number finished in the product matrix; you'll have to follow the same process five more times to get the five other elements in the final answer. For instance, to calculate the element p_{21}, place your left index finger on element 1 (the leftmost number in row *two* of the left matrix) and your right index finger on element –4 (the topmost number in column *one* of the right matrix). As you move your left finger horizontally and your right finger vertically (multiplying pairs of numbers as you go), you'll get

$$p_{21} = 1(-4) + 7(5)$$

$$p_{21} = -4 + 35$$

$$p_{21} = 31$$

Here are the results you should get for every element in the matrix, written in matrix form:

$$\begin{bmatrix} 2 & -2 \\ 1 & 7 \end{bmatrix} \cdot \begin{bmatrix} -4 & 0 & -9 \\ 5 & 3 & -6 \end{bmatrix} = \begin{bmatrix} 2(-4)+(-2)(5) & 2(0)+(-2)(3) & 2(-9)+(-2)(-6) \\ 1(-4)+7(5) & 1(0)+7(3) & 1(-9)+7(-6) \end{bmatrix}$$
$$= \begin{bmatrix} -8-10 & 0-6 & -18+12 \\ -4+35 & 0+21 & -9-42 \end{bmatrix}$$
$$= \begin{bmatrix} -18 & -6 & -6 \\ 31 & 21 & -51 \end{bmatrix}$$

You've Got Problems

Problem 4: Calculate the product.

$$\begin{bmatrix} 2 & -5 \\ -3 & 1 \\ -4 & 0 \end{bmatrix} \cdot \begin{bmatrix} 6 & -2 \\ -1 & 3 \end{bmatrix}$$

Determining Determinants

A *determinant* is a real number value that is defined for square matrices only. For now, don't ask yourself, "Why should I find a determinant?" (I'll show you one practical application for them in the final section of this chapter.) Instead, focus on "*How* do I find a determinant?" Also, if you're getting frustrated, don't ask yourself, "Why *should* I learn algebra?"; instead, ask yourself, "How in the world did the opposite sex *ever* find me attractive until I decided to compute matrix products?" It's a better ego boost, and mathematicians use it all the time. Viva self-delusion!

Talk the Talk

Every square matrix has a real number associated with it called a **determinant;** you will use it to solve systems of linear equations later on in the chapter.

Even though all square matrices have determinants, in algebra you are generally only required to calculate determinants for 2×2 and 3×3 matrices. Once the dimensions get larger, the process becomes markedly more difficult, so it's usually discussed in more advanced courses, such as precalculus.

To signify that you're calculating a determinant for a matrix, draw thin lines (which look a lot like, but definitely are not, absolute value bars) on either side of the matrix, rather than the typical brackets. For example, to indicate the determinant of the matrix $A = \begin{bmatrix} 4 & 6 \\ -2 & -5 \end{bmatrix}$, you could write either $|A|$ or $\begin{vmatrix} 4 & 6 \\ -2 & -5 \end{vmatrix}$.

2 × 2 Determinants

The determinant of a 2 × 2 matrix $\begin{vmatrix} a & b \\ c & d \end{vmatrix}$ is equal to $ad - cb$. In other words, multiply the numbers diagonal to one another and then subtract, as illustrated in Figure 9.2.

$$\begin{vmatrix} a & b \\ c & d \end{vmatrix} = ad - cb$$

Figure 9.2

Subtract the products of the diagonal numbers to get the determinant of a 2 × 2 matrix.

Example 5: If $A = \begin{bmatrix} 4 & 6 \\ -2 & -5 \end{bmatrix}$, calculate $|A|$.

Solution: Picture the diagonal lines from Figure 9.2, which direct you to multiply 4(–5) and then subtract –2(6).

$$|A| = 4(-5) - (-2)(6)$$
$$|A| = -20 - (-12)$$
$$|A| = -20 + 12$$
$$|A| = -8$$

Kelley's Cautions

Make sure that you subtract in the correct order when calculating a 2 × 2 determinant, or you'll get the sign of your answer wrong. Subtract the product of the upward diagonals.

Don't forget that subtraction sign between 4(–5) and –2(6) required by the formula!

You've Got Problems

Problem 5: Evaluate $\begin{vmatrix} 9 & -4 \\ 3 & -1 \end{vmatrix}$.

3×3 Determinants

The process necessary to calculate a 3×3 determinant is a little different than the one you used for the 2×2 variety. Basically, you're still multiplying along diagonals and then subtracting. In fact, you're still subtracting all diagonals that point in the upper-right direction, just like you did before.

However, to use the simple diagonal method, you'll have to rewrite the matrix first. Here are the steps to follow in order to calculate the determinant of the 3×3 matrix

$$\begin{bmatrix} a & b & c \\ d & e & f \\ g & h & i \end{bmatrix}$$

1. **Copy the first two columns of the matrix.** Rewrite those first two columns to the immediate right of the matrix's third column. The final result should be a 3×5 matrix whose first and fourth columns match and whose second and fifth columns match.
2. **Multiply along the diagonals.** In 2×2 determinants, you multiplied diagonally down and then subtracted the upward diagonal. Similarly, in 3×3 matrices, you multiply diagonally down three times (adding the results) and then subtract three upward diagonals, as pictured in Figure 9.3.

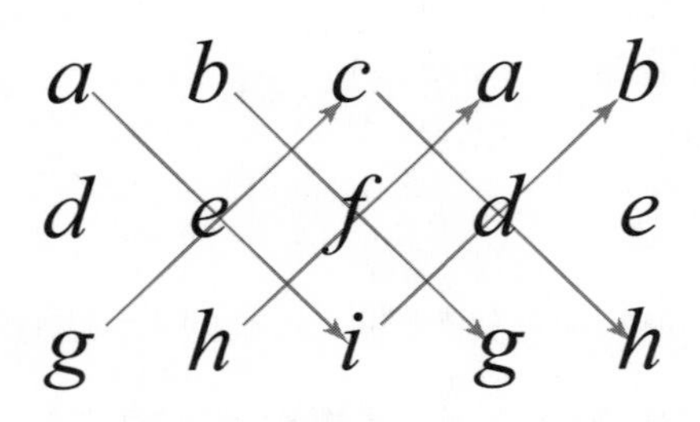

Figure 9.3

Multiply along these diagonals to calculate the determinant of a 3×3 matrix. Make sure to subtract the product of each diagonal that points upward.

Critical Point

Most graphing calculators allow you to input matrices and then will calculate determinants for you. This is a good way to check your work during homework, until you feel confident calculating them by hand.

If you multiply as Figure 9.3 indicates, the determinant of the matrix will be

$$|A| = aei + bfg + cdh - gec - hfa - idb$$

Example 6: Calculate the determinant.

$$\begin{vmatrix} -3 & 5 & 1 \\ 2 & -6 & 0 \\ -7 & 4 & -1 \end{vmatrix}$$

Solution: Copy the first two columns to the right of the original matrix.

$$\begin{matrix} -3 & 5 & 1 & -3 & 5 \\ 2 & -6 & 0 & 2 & -6 \\ -7 & 4 & -1 & -7 & 4 \end{matrix}$$

Multiply along the diagonals pictured in Figure 9.3, remembering to put a negative sign in front of every upward-pointing diagonal.

$$\begin{aligned} &(-3)(-6)(-1)+(5)(0)(-7)+(1)(2)(4)-(-7)(-6)(1)-(4)(0)(-3)-(-1)(2)(5) \\ &=(-18)+(0)+(8)-(42)-(0)-(-10) \\ &=-18+0+8-42-0+10 \\ &=-42 \end{aligned}$$

You've Got Problems

Problem 6: Calculate the determinant.

$$\begin{vmatrix} -2 & -4 & 9 \\ 3 & -5 & 1 \\ 7 & 2 & -3 \end{vmatrix}$$

Cracking Cramer's Rule

Cramer's Rule is a neat trick that solves systems of equations, although I'd be lying if I called it a shortcut. Quite honestly, the elimination method you learned in Chapter 8 solves systems quickly enough, but Cramer's Rule is the simplest procedure that makes use of determinants, so it's included in most contemporary algebra courses.

Since most algebra students spend a lot of the year wondering "Why do I have to learn how to do any of this?" Cramer's Rule gives the teacher the rare opportunity to show why learning a strange skill (like calculating determinants) can be useful. True, solving systems of equations is not all that useful in day-to-day life ("Uh oh, my canoe just rolled over in the middle of the river; good thing I know how to solve systems of equations, or I'd never make it to shore ...."); at least it's something concrete in the maddeningly abstract world of algebra concepts.

Let's say you're trying to solve the system of equations

$$\begin{cases} ax+by=c \\ dx+ey=f \end{cases}$$

using Cramer's Rule. (Note that both of the equations in the system are in standard form.) Here are the steps you'd follow:

1. **Write the coefficient matrix.** The coefficient matrix, C, is the 2×2 matrix containing the x-coefficients in the first column and the y-coefficients in the second.

$$C = \begin{bmatrix} a & b \\ d & e \end{bmatrix}$$

2. **Replace the columns in C one at a time.** You'll design two new 2×2 matrices, A and B, which are both based on the coefficient matrix C. To get matrix A, start with the coefficient matrix, but replace the first column with the constants from the original system of equations (the numbers c and f from the right sides of the equal signs). In other words, instead of a and d in the first column of A, you'll have c and f. In matrix B, you'll replace the *second* column of the coefficient matrix with the constants.

$$A = \begin{bmatrix} c & b \\ f & e \end{bmatrix} \qquad B = \begin{bmatrix} a & c \\ d & f \end{bmatrix}$$

3. **Calculate the determinants of the matrices.** To figure out the solution to the system, you'll need to know $|A|$, $|B|$, and $|C|$.

4. **Use the Cramer's Rule formulas.** The solution to the system, if it exists, will be a set of coordinate pairs (x,y). To figure out x and y, just plug in the determinants you just evaluated into these formulas:

$$x = \frac{|A|}{|C|}, \quad y = \frac{|B|}{|C|}$$

> CAUTION
>
> **Kelley's Cautions**
>
> If the system is inconsistent or dependent, $|C|$ will equal 0, meaning that Cramer's method will not work. (You wind up with a 0 in the denominator of $x = \frac{|A|}{|C|}$ and $y = \frac{|B|}{|C|}$.)

It seems like there are a lot of steps, but most of them are very quick and painless, making Cramer's Rule a viable option for solving systems once you practice long enough to fly through the steps quickly.

Example 7: Solve the system using Cramer's Rule.

$$\begin{cases} 4x + 6y = -9 \\ 3x - 8y = -13 \end{cases}$$

Solution: First, create the coefficient matrix C, using the coefficients of x and y:

$$C = \begin{bmatrix} 4 & 6 \\ 3 & -8 \end{bmatrix}$$

How'd You Do That?

You can also apply Cramer's Rule to systems of three equations in three variables. Let's say you're solving this system via Cramer's Rule:

$$\begin{cases} ax + by + cz = d \\ ex + fy + gz = h \\ jx + ny + pz = q \end{cases}$$

Create the coefficient matrix C like before, but realize that it will now be 3 × 3; since there are three variables, there will be three columns, representing the coefficients of the x's, y's and z's, respectively.

Design matrices A and B by replacing the first and second columns, respectively, with the column of constants, just like you did when there were only two equations. However, since these matrices are 3 × 3, you'll need to create a fourth matrix, D, in which the third column of the coefficient matrix is replaced by the constants.

The solution to the system will be:

$$x = \frac{|A|}{|C|},\quad y = \frac{|B|}{|C|},\quad z = \frac{|D|}{|C|}$$

To create matrix A, replace the first column with the column of constants, the numbers –9 and –13. To create B, start with C and replace the second column with the constants.

$$A = \begin{bmatrix} -9 & 6 \\ -13 & -8 \end{bmatrix} \quad B = \begin{bmatrix} 4 & -9 \\ 3 & -13 \end{bmatrix}$$

Now calculate the determinants of all the matrices.

$$|A| = (-9)(-8) - (-13)(6) = 72 + 78 = 150$$
$$|B| = (4)(-13) - (3)(-9) = -52 + 27 = -25$$
$$|C| = (4)(-8) - (3)(6) = -32 - 18 = -50$$

You're nearly finished. Just plug these values into the Cramer's Rule formula.

$$x = \frac{|A|}{|C|},\quad y = \frac{|B|}{|C|}$$
$$x = \frac{150}{-50},\quad y = \frac{-25}{-50}$$
$$x = -3,\quad y = \frac{1}{2}$$

You've Got Problems

Problem 7: Solve the system using Cramer's Rule.

$$\begin{cases} 6x - 5y = -7 \\ 3x + 2y = 1 \end{cases}$$

The Least You Need to Know

- Matrices are orderly collections of numbers, organized in rows and columns.
- In order for the matrix product $A \cdot B$ to exist, the number of columns in A must equal the number of rows in B.
- Determinants are defined for all square matrices.
- Cramer's Rule is a technique that makes use of determinants to solve systems of equations.

Part 4 Now You're Playing with (Exponential) Power!

It's time for an amicable breakup from linear equations. You should meet together at a local restaurant (so the lines won't cause a scene) and give them the old "It's not you, it's me" speech. You need more out of life than just slopes and intercepts. You need a graph that's a little more unpredictable, a graph with a little curve to it. In this part, you'll enter a strange new world of equations whose variables contain exponents. Sometimes it will feel a little strange, and you'll yearn for the days of your first linear love, but trust me, this is best for both of you.

Chapter 10

Introducing Polynomials

In This Chapter

- Classifying polynomials
- Adding, subtracting, and multiplying polynomials
- Calculating polynomial quotients via long division
- Performing synthetic division

So far, most of the simplifying you've had to do has been pretty basic. For example, in Chapter 3, you learned exponential laws, which taught you that the product $(2x^4)(3x^7)$ is equal to $2 \cdot 3 \cdot x^{4+7} = 6x^{11}$. In Chapter 4, when you learned to solve equations, you may have manipulated the equation $3x = 9 + 2x$ by subtracting $2x$ from both sides to get $x = 9$. What you didn't know was that in both cases, you were actually simplifying polynomial expressions.

I haven't made a big deal about defining polynomials or gone into a lot of detail about the mechanics of simplifying such expressions so far, mainly because the sorts of things you've done to this point feel good intuitively—they sort of make sense. Sure, $5x + 9x$ should equal $14x$; that feels right. However, before you go any farther, I want to tie up any loose ends and get specific about what kinds of things you can and can't add together, and even explore some complicated topics, like dividing variable expressions.

Classifying Polynomials

A *polynomial* is basically a string of mathematical clumps (called *terms*) all added together. Each individual clump usually consists of one or more variables raised to exponential powers, usually with a coefficient attached. Polynomials can be as simple as the expression $4x$, or as complicated as the expression $4x^3 + 3x^2 - 9x + 6$.

Polynomials are usually written in standard form, which means that the terms are listed in order from the largest exponential value to the term with the smallest exponent. Because the term containing the variable raised to the highest power is listed first in standard form, its coefficient is called the *leading coefficient.* A polynomial not containing a variable is called the *constant.*

For example, if you were to write the polynomial $2x^3 - 7x^5 + 8x + 1$ in standard form, it would look like this: $-7x^5 + 2x^3 + 8x + 1$. (Note that each term's variable has a lower power than the term to its immediate left.) The *degree* of this polynomial is 5, its leading coefficient is –7, and the constant is 1.

Talk the Talk

A **polynomial** consists of the sum of distinct algebraic clumps (called **terms**), each of which consists of a number, one or more variables raised to an exponent, or both. The largest exponent in the polynomial is called the **degree,** and the coefficient of the variable raised to that exponent is called the **leading coefficient.** The **constant** in a polynomial has no variable written next to it.

Technically, the constant in a polynomial *does* have a variable attached to it, but the variable is raised to the 0 power. For example, you could rewrite the simple polynomial $2x + 1$ as $2x + 1x^0$, but since $x^0 = 1$ (and anything multiplied by 1 equals itself), there's no reason to write x^0 at the end of the polynomial.

Because there are so many different kinds of polynomials (52 flavors at last check, including pistachio), there are two techniques that are used to classify them, one based on the number of terms a polynomial contains (see Table 10.1), and one based on the degree of the polynomial (see Table 10.2).

Table 10.1 Classifying a Polynomial Based on the Number of Its Terms

Number of Terms	Classification	Example
1	monomial	$19x^2$
2	binomial	$3x^3 - 7x^2$
3	trinomial	$2x^2 + 5x - 1$

Notice that there are only special classifications for polynomials according to the number of their terms if that number is three or less. Polynomials with four or more terms are either classified according to degree or just described with the ultra-generic (and not very helpful) label "polynomial." (It's just as specific as labeling you a "human being.")

Table 10.2 Classifying a Polynomial Based on Its Degree

Degree	Classification	Example
0	constant	$2x^0$ or 2
1	linear	$6x^1 + 9$ or $6x + 9$
2	quadratic	$4x^2 - 25x + 6$
3	cubic	$x^3 - 1$
4	quartic	$2x^4 - 3x^2 + x - 8$
5	quintic	$3x^5 - 7x^3 - 2$

There are more degree classifications for polynomials, but those listed in Table 10.2 are by far the most commonly used.

When classifying a polynomial, you don't have to choose one method or the other. In fact, if you classify the polynomial both ways at once, whenever possible, you paint a more descriptive picture of it.

Example 1: Classify the following polynomials.

(a) $3 - 4x - 6x^2$

Solution: This polynomial has three terms, so it's a trinomial. Furthermore, its degree is 2, which makes it quadratic. So, all together, it's a quadratic trinomial. When you use both classifications at once, write the degree classifier first since it's an adjective ("trinomial quadratic" just doesn't sound right).

Critical Point

If you're asked to classify a polynomial like $3x^3y^2 - 4xy^3 + 6x$ (which contains more than one kind of variable in some or all of its terms) according to its degree, add the exponents in each term together. The highest total will be the degree. In $3x^3y^2 - 4xy^3 + 6x$, the degree is 5, since the highest exponent total comes from the first term, and $3 + 2 = 5$.

(b) 13

Solution: There's only one term, and it has no variable written explicitly; therefore, this is the same thing as $13x^0$. This expression is best classified as a constant monomial.

You've Got Problems

Problem 1: Classify the following polynomials:

(a) $4x^3 + 2$

Adding and Subtracting Polynomials

In previous chapters, you've simplified expressions like $3x + 7x$ to get $10x$, or perhaps subtracted terms like $5y - 9y$ to get $-4y$. That arithmetic makes perfect sense if you translate the mathematics into words. For example, the expression $3x + 7x$ literally means "three of a certain number added to seven more of that same number," which is definitely equal to "10 of that number," or $10x$.

Talk the Talk

Like terms have variables which match exactly, like $4x^2y^3$ and $-7x^2y^3$. You can only add or subtract two terms if they are like terms.

What I didn't tell you back then was that you were only allowed to combine the coefficients of those terms because they contained *the exact same variables.* Actually, any two terms whose variables match *exactly* are called *like terms*, and you cannot add terms together or subtract them from one another if they are not like terms.

Kelley's Cautions

Many students try to simplify the expression $4x + 5y$ and get $9xy$, but that's wrong! Remember, you can't add or subtract $4x$ and $5y$ because the terms have different variables. It would be like adding four cats to five dogs and getting nine dats (or cogs). Unlike terms are like apples and oranges—you can't combine them.

If two terms have the same variables and get all nervous when they look at each other, you can upgrade them from like terms to love terms, but since it's hard to read the emotions of variables (they're always changing on you), most mathematicians don't even try to make that differentiation.

Now you know why the expression $13x^2y^3 - 5x^2y^3$ can be simplified as $8x^2y^3$. Since the variables in both terms match exactly (they both contain x^2y^3), all you have to do is combine the coefficients and attach a copy of the matching variable string.

Example 2: Simplify the following expression.

$$4x^3 + 5x^2 - 3x + 1 - (2x^3 - 8x^2 + 9x - 6)$$

Solution: Start by applying the distributive property (multiply everything in the parentheses by –1).

$$4x^3 + 5x^2 - 3x + 1 - 2x^3 + 8x^2 - 9x + 6$$

If you rewrite the expression so that like terms are grouped together, it makes simplifying easier.

$$4x^3 - 2x^3 + 5x^2 + 8x^2 - 3x - 9x + 1 + 6$$

Combine the coefficients of each pair of like terms.

$$2x^3 + 13x^2 - 12x + 7$$

You've Got Problems

Problem 2: Simplify the expression $3x^2 - 9 + 2(x^2 - 6x + 5)$.

Multiplying Polynomials

Unlike addition and subtraction, you don't need like terms in order to multiply polynomials (nor do you need like terms to divide polynomials, but I'll discuss that in the next section). In fact, multiplying polynomials is actually pretty easy. All you have to do is apply exponential rules and the distributive property, both of which you learned in Chapter 3.

Products of Monomials

Here's what you should do to multiply two monomials together:

1. **Multiply their coefficients.** The result is the coefficient of the answer.
2. **List all the variables that appear in either term.** These should follow the coefficient you got in step 1, preferably in alphabetical order.
3. **Add up the powers.** Determine the sums of matching variables' exponents and write them above the corresponding variable in the answer.

How'd You Do That?

Step 3 tells you to add the powers of matching variables because of the exponential rule from Chapter 3 stipulating that $x^a \cdot x^b = x^{a+b}$. (The product of exponential expressions with matching bases equals the base raised to the sum of the powers.)

Even if the steps seem weird at first, don't worry. Multiplying monomials is a skill you'll understand very quickly.

Example 3: Calculate the products.

(a) $(-3x^2y^3z^5)(7xz^3)$

Solution: First multiply the coefficients: $-3 \cdot 7 = -21$; then, list all the variables that appear in the problem in alphabetical order. (It doesn't matter that the second monomial doesn't contain a *y*. As long as a variable appears anywhere in the problem, you should list it next to the coefficient you just found.)

$$-21xyz$$

Add up the exponents for each variable you listed. The first term has *x* to the 2 power, and the second term has *x* to the 1 power, so the answer will have *x* to the 2 + 1 = 3 power. Similarly, the *z* power of the answer should be 8, since there's a *z* to the 5 power in the first monomial and a *z* to the 3 in the second. Since there's only one *y* term, you just copy its power to the final answer; there's nothing to add.

$$-21x^3y^3z^8$$

(b) $3w^2x(2wxy - x^2y^2)$

Solution: Apply the distributive property, multiplying both terms by $3w^2x$.

$$3w^2x(2wxy) + 3w^2x(-x^2y^2)$$

Find each product separately.

$$6w^3x^2y - 3w^2x^3y^2$$

You've Got Problems

Problem 3: Calculate the product.

$$3x^2y(5x^3 + 4x^2y - 2y^5)$$

Binomials, Trinomials, and Beyond

Calculating polynomial products is kind of freeing. As I've said, two terms need not have anything in common to be multiplied together. (Based on couples I've met, I think the same is true for people, but I digress.) However, so far you can only multiply polynomial expressions if one of them is a monomial. In Example 3(a), you had two monomials, and in Example 3(b) and Problem 3, you were distributing a monomial. It turns out that multiplying polynomials with more than one term can be accomplished through a slightly modified version of the distributive property.

Thanks to the distributive property, you already know that the expression $a(b + c)$ can be rewritten as $ab + ac$; just multiply the a by each thing in the parentheses. In a similar fashion, you can calculate the product of the expression $(a + b)(c + d)$, even though in this case, you're multiplying binomials. Instead of just distributing a, like you did moments ago, you'll distribute each term in the first binomial through the second binomial, one at a time.

In other words, you'll multiply everything in the second binomial by a and then go through and do it again, this time multiplying everything by b.

$$ac + ad + bc + bd$$

Critical Point

Some algebra teachers focus on the FOIL method, a technique for multiplying two binomials. Each letter stands for a pair of terms in the binomials, the first, outside, inside, and last terms.

If you've never heard of FOIL, that's fine, because it only works for the special case of multiplying two binomials, whereas my multiple distribution technique works for all polynomial products. Besides, if you use my method, you actually end up doing FOIL anyway, even though it's unintentional.

So, you're still distributing, you're just doing it twice, that's all. What if you were multiplying a trinomial by a trinomial? Follow the same procedure; distribute each term in the first polynomial through the second, one at a time.

$$(a + b + c)(d + e + f) = ad + ae + af + bd + be + bf + cd + ce + cf$$

In case you're wondering, the numbers of terms in the polynomials don't have to match. You could multiply a binomial times a trinomial just as easily, as you'll see in Example 4.

Example 4: Find the product and simplify.

$$(x - 2y)(x^2 + 2xy - y^2)$$

Solution: Each term of the left polynomial, x and $-2y$, should be distributed through the second polynomial, one at a time.

$$(x)(x^2) + (x)(2xy) + (x)(-y^2) + (-2y)(x^2) + (-2y)(2xy) + (-2y)(-y^2)$$

If you place all of the terms in parentheses, you don't have to worry about signs right away. It doesn't matter if some terms are positive and some are negative; just write them all in parentheses and add all the products together.

Kelley's Cautions

Once you multiply, always make sure to see if you can simplify the result. Just about every algebra teacher in the world demands simplified answers, and if you don't comply, they've been known to do things like mark answers wrong, take points off, or (in extreme cases) get so angry that they send a cybernetic organism back in time to kill you before you sign up for their class.

Now all you have to do is multiply pairs of monomials together.

$$x^3 + 2x^2y - xy^2 - 2x^2y - 4xy^2 + 2y^3$$

The directions for the problem tell you to simplify, which means you should now look for like terms which can be combined. If you look closely, you'll see that the terms $2x^2y$ and $-2x^2y$ have the same variable, so they can be combined to get 0 (they're opposites of one another, so they'll cancel each other out). In addition, you can combine the terms $-xy^2$ and $-4xy^2$ to get $-5xy^2$.

$$x^3 - 5xy^2 + 2y^3$$

You've Got Problems

Problem 4: Find the product and simplify.

$$(2x + y)(x - 3y)$$

Dividing Polynomials

There are two techniques you can use to calculate the quotient of two polynomials, one (which may feel a bit familiar) will work for all polynomial division problems but takes a while, whereas the other will work much faster, but only works in specific circumstances.

Talk the Talk

In the division problem $a \div b$ (which is typically rewritten as $b\overline{)a}$ when you begin calculating the answer), b is called the **divisor** and a is the **dividend**.

Long Division

The most reliable way to divide polynomials is the process of long division; even though the process is a bit cumbersome, it works for every division problem you'll see. It actually replicates the technique you learned in elementary school to divide whole numbers, in case you're wondering why it feels familiar.

Example 5: Calculate the quotient of $(x^3 + 5x^2 - 3x + 4) \div (x^2 + 1)$.

Solution: Start by rewriting the problem in long division notation; the *divisor* (what you're dividing by) goes to the left, and the *dividend* (what you're dividing into) is written beneath the symbol. As you're writing the polynomials, make sure there are no missing exponents in either the divisor or dividend.

In this problem, the divisor has no x term; you don't want it to be missing, so you should write it in there with a coefficient of 0. (If you don't, things won't line up right.)

$$x^2 + 0x + 1 \overline{)\, x^3 + 5x^2 - 3x + 4}$$

Look at the first term of the divisor, x^2, and the first term of the dividend, x^3. Ask yourself, "What times x^2 will give me *exactly* x^3?" The answer is x, and you should write that answer above the division symbol. In fact, you should write it directly above $-3x$, since it and the number you just came up with (x) are like terms.

$$\begin{array}{r} x \\ x^2 + 0x + 1 \overline{)\, x^3 + 5x^2 - 3x + 4} \end{array}$$

Multiply that x and the divisor together, and write the result below the dividend, so that the like terms line up; draw a horizontal line beneath the product.

$$\begin{array}{r} x \\ x^2 + 0x + 1 \overline{)\, x^3 + 5x^2 - 3x + 4} \\ \underline{x^3 + 0x^2 \quad + x} \end{array}$$

Now multiply everything in that bottom line by -1 and then combine the result with the like terms directly above. Write the result below the horizontal line.

$$\begin{array}{r} x \\ x^2 + 0x + 1 \overline{)\, x^3 + 5x^2 - 3x + 4} \\ \underline{-x^3 - 0x^2 \quad - x} \\ 5x^2 - 4x \end{array}$$

> **How'd You Do That?**
>
> The reason you ask yourself "What times x^2 will give me *exactly* x^3?" is so that once you write the answer, multiply it by the divisor, and multiply it by -1, you'll get the exact opposite of the dividend's first term. That way, when you combine terms, you'll get $x^3 - x^3$, which equals 0, eliminating a term.

Drop down the next term in the dividend polynomial, in this case a positive 4, and repeat the above process, beginning with the question, "What times x^2 (the first term in the divisor) will give me exactly $5x^2$ (the first term in the subtraction problem you just finished); the answer is 5. Write that constant above the division symbol (right above 4, its like term), multiply it times the divisor, change each of the terms in the product to its opposite, and combine the like terms.

$$\begin{array}{r} x+5 \\ x^2+0x+1\overline{)\ x^3+5x^2-3x+4} \\ \underline{-x^3-0x^2\ \ -x\quad} \\ 5x^2-4x+4 \\ \underline{-5x^2-0x-5} \\ -4x-1 \end{array}$$

If there had been additional terms in the dividend, you'd drop them down one at a time and repeat the whole process again, but since there are no more terms to drop, you're finished. The quotient is the quantity above the division symbol, $x + 5$, and the remainder is $-4x - 1$.

You should write your answer as the quotient plus the fraction whose numerator is the remainder and whose denominator is the original divisor, like so:

$$x+5+\frac{-4x-1}{x^2+1}$$

You can check your answer by multiplying the quotient $(x + 5)$ times the original divisor $(x^2 + 1)$ and then adding the remainder.

$$\begin{aligned} &(\text{quotient})(\text{divisor})+(\text{remainder}) \\ &=(x+5)(x^2+1)+(-4x-1) \\ &=x^3+x+5x^2+5-4x-1 \\ &=x^3+5x^2+x-4x+5-1 \\ &=x^3+5x^2-3x+4 \end{aligned}$$

If you did everything correctly, you should get the original dividend; that's exactly what happened here, so you can bask in the glow of your own mathematical greatness, secure in the knowledge that you rule.

You've Got Problems

Problem 5: Calculate the quotient of $(x^2 - 7x + 8) \div (x + 4)$.

Synthetic Division

If you're trying to divide a polynomial by a linear binomial (in the form "$x - c$" where c could be any real number), then the best way to do it is through synthetic division.

For example, the division problem $(x^3 - 2x^2 + 3 - 4) \div (x + 3)$ would be a good candidate for synthetic division; the divisor is technically in the form $x - c$, because if you set $c = -3$, then $x - c = x - (-3) = x + 3$. However, the problem $(x^3 - 2x^2 + 3x - 4) \div (x^2 + 3)$ would not be a good candidate, since the divisor is not linear.

Synthetic division is much simpler than long division because all you use are the coefficients of the polynomials. I don't know why it's called "synthetic" division—it's not unnatural, filled with preservatives, or fake. It's never been to the plastic surgeon to get a nip, a tuck, an implant, or a reduction, so the name puzzles me. Like long division, the best way to learn the process is through an example, so I'll get right to it.

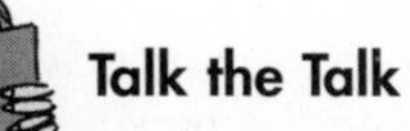

Talk the Talk

Synthetic division is a shortcut technique for calculating polynomial quotients that's only applicable when the divisor is of the form $x - c$, where c is a real number.

Example 6: Calculate the quotient of $(2x^3 - x + 4) \div (x + 3)$.

Solution: First check to see if there are any missing powers of x in the dividend; notice that there's no x^2 power in there, so insert it with a coefficient of 0 (just like you did in long division) to get a dividend of $2x^3 + 0x^2 - x + 4$. Now list those coefficients in descending order of their exponents.

$$2 \quad 0 \quad -1 \quad 4$$

To the left of that list, write the *opposite* of the constant in the divisor. In this problem, the divisor is $x + 3$, so the opposite of its constant will be -3. It's separated from the rest of the coefficients with something that looks like a half-box. Leave some space beneath that row and draw a horizontal line.

$$\begin{array}{r|rrrr} -3 & 2 & 0 & -1 & 4 \\ & & & & \\ \hline & & & & \end{array}$$

The setup is all finished, and it's time to get started. Take the leading coefficient (2) and drop it below the horizontal line.

$$\begin{array}{r|rrrr} -3 & 2 & 0 & -1 & 4 \\ & & & & \\ \hline & 2 & & & \end{array}$$

Critical Point

Even though synthetic division can only be applied in a specific situation (given a linear divisor of the form $x - c$), it will be extremely useful later on in Chapter 14. So, even though it's basically a shortcut to long division, it's definitely worth learning.

Multiply the number in the half-box (–3) times the number below the line (2) and write the result (–6) below the next coefficient (0).

$$\begin{array}{r|rrrr} -3 & 2 & 0 & -1 & 4 \\ & & -6 & & \\ \hline & 2 & & & \end{array}$$

Combine the numbers in the second column (0 – 6 = –6) and write the result directly below the numbers you just combined.

$$\begin{array}{r|rrrr} -3 & 2 & 0 & -1 & 4 \\ & & -6 & & \\ \hline & 2 & -6 & & \end{array}$$

Repeat the process two more times, each time multiplying the number in the box by the new number below the line, writing the result in the next column, adding the numbers in that column together, and writing the result below the line again.

$$\begin{array}{r|rrrr} -3 & 2 & 0 & -1 & 4 \\ & & -6 & 18 & -51 \\ \hline & 2 & -6 & 17 & -47 \end{array}$$

The answer will be each of the numbers below the line with decreasing powers of x next to them; start with the x power one less than the degree of the dividend. Therefore, the first power of x in your answer for this problem should be 2. In case you're wondering, the rightmost number below the line is the remainder, which is written just like it was in long division.

$$2x^2 - 6x + 17 + \frac{-47}{x+3}$$

You can check this the same way you did long division: $(2x^2 - 6x + 17)(x + 3) - 47$ should equal $2x^3 - x + 4$, and it does.

You've Got Problems

Problem 6: Calculate the quotient of $(4x^3 - 2x^2 - 10x + 1) \div (x - 2)$.

The Least You Need to Know

- If two terms in a polynomial have variables that match exactly, they are like terms.
- Only like terms can be added or subtracted.
- Multiply polynomials by distributing terms one at a time.
- Long division will correctly calculate polynomial quotients for any problem, but synthetic division (which only works for special binomial divisors) is faster.

Chapter 11

Factoring Polynomials

In This Chapter

- Finding greatest common factors
- Recognizing famous factoring patterns
- Factoring trinomials via their coefficients
- Factoring difficult trinomials with the bomb technique

I remember the first day I tried to drive in reverse. I didn't think it would be significantly harder (after all, walking backwards is no harder than walking forwards, so why should driving be any different?), but boy, was I wrong. Everything feels different when you're backing up. No matter which way you turn the wheel, the car seems to go in the opposite direction you intend. I still get a little antsy in reverse sometimes; it's a whole new ballgame.

Now that you are used to multiplying polynomials (in my metaphor, the equivalent of driving), now it's time to throw the whole process in reverse and learn factoring.

Why is factoring the reverse of multiplication, you ask? It's because *factoring* is the process of taking what was once a product and breaking it into the original pieces (called *factors*) that multiply together to get that product.

In Chapter 10, you learned to calculate $(x - 3)(x + 5) = x^2 + 2x - 15$. In this chapter, you'll start with $x^2 + 2x - 15$ and wind up with $(x - 3)(x + 5)$. At first, things will feel a little strange, like driving in reverse originally does, but soon you'll master it as well.

Greatest Common Factors

The *greatest common factor* (GCF) of a polynomial is the largest monomial that divides evenly into each term. It's very similar to the greatest common factor you calculated in Chapter 2, except that polynomial GCFs usually contain one or more variables.

Talk the Talk

Factoring is the process of returning a polynomial product back to its original, unmultiplied pieces, called **factors**. The simplest technique for factoring involves identifying a polynomial's **greatest common factor**, the largest monomial that divides evenly into each of the polynomial's terms.

Here's how to calculate the GCF of a polynomial:

1. **Find the GCF of the polynomial's coefficients.** This will be the coefficient of the polynomial's GCF.
2. **Identify common variable powers.** Look at the variables in each term of the polynomial. The GCF should contain the highest possible power of every variable. Here's the catch: Every term must contain the variable raised to *at least* that exponent.
3. **Multiply.** The product of steps 1 and 2 above is the GCF of the polynomial.

Once you've found the GCF of the polynomial, you can factor that polynomial. Just write the GCF followed by a set of parentheses. Inside those parentheses, you should list what's left of each polynomial term once you divide it by the GCF. In other words, the parentheses show the polynomial with the GCF "sucked out."

Example 1: Factor the polynomial $6x^2y^3 - 12xy^2$.

Solution: Start by finding the GCF of the polynomial. Its coefficient will be 6, the GCF of 6 and 12. To determine its variable part, ask yourself, "What is the maximum number of each variable that's contained in *every* term?" (If that doesn't work, ask yourself, "Why did I ever try to figure out algebra, anyway? It's sucking away my will to live!" and flail your arms madly in the air. It won't help you solve the problem, but you'll definitely feel better.) Look at the x's; the first term is squared, so it has two of them, but the second term only has one. Therefore, the largest number contained by both is 1, and the GCF will contain x to the power of 1.

On the other hand, both terms contain at least two y's, so the GCF will also contain a y^2. Put all three pieces together to get a GCF of $6xy^2$. Now, divide every term by the GCF.

$$\frac{6x^2y^3}{6xy^2} = xy \qquad \frac{-12xy^2}{6xy^2} = -2$$

You don't have to use long division to get those answers. Start by dividing the coefficients. In the first term, $6 \div 6 = 1$, and in the second, $-12 \div 6 = -2$. Then, apply the exponential law that states $\frac{x^a}{x^b} = x^{a-b}$ to each term. (Subtract the denominator power from the numerator power for each matching variable.) For instance, in the first term, you'll get $x^{2-1} = x^1 = x$ and $y^{3-2} = y^1 = y$.

You're almost done. The factored form of the original polynomial will equal the GCF times the divided form of the terms you just calculated. Just write the divided form of the terms inside parentheses and multiply that entire quantity by the GCF.

$$6xy^2(xy - 2)$$

It's easy to check your answer. Just distribute the $6xy^2$ term through the parentheses, and you should end up with the original problem.

You've Got Problems

Problem 1: Factor the polynomial $9x^5y^2 + 3x^4y^3 - 6x^3y^7$.

Factoring by Grouping

Factoring by grouping is a technique used in a very specific case, and is likely the technique you'll use the least in this chapter. However, when it's applicable, it gets the job done, and fast.

It works best when you're given a polynomial whose terms don't *all* have a greatest common factor, but perhaps *some* of them do have factors in common. Basically, you'll group the large polynomial into two smaller parts, both of which contain terms with common factors. Then, you'll factor both of those smaller groups individually. Sound confusing? Probably. However, when you see it at work in an example, it's easier to understand what's going on.

Example 2: Factor the polynomial $2x^3 - 4x^2 - 3x + 6$.

Solution: Unfortunately, these terms don't have any factor in common (except 1, and that's true of any group of terms). However, look at the polynomial as two groups, each containing two terms.

$$(2x^3 - 4x^2) + (-3x + 6)$$

Notice that the left group has a GCF of $2x^2$; go ahead and factor that out and pay special attention to the binomial that results.

$$2x^2(x - 2) + (-3x + 6)$$

You can factor a 3 out of the second group of terms, but it's more useful to factor out a –3. Why? If you do, the binomial that results *exactly matches* the binomial you got when you factored the first group.

$$2x^2(x - 2) - 3(x - 2)$$

Here's the part that sometimes confuses students. You should now factor out the binomial $(x - 2)$ from both terms. At first it feels weird factoring out something other than a monomial, but it's definitely allowed. When you pull that $(x - 2)$ out, all you're left with in the first term is the $2x^2$, which was hanging out front, and in the second term, just the –3 remains. So, write the binomial, and pop the leftovers in a set of parentheses after it, like so:

$$(x - 2)(2x^2 - 3)$$

CAUTION

Kelley's Cautions

The key to ensuring that factoring by grouping works right is getting the factored binomials to match; in Example 2, both binomials were $(x - 2)$. Basically, this means you should decide whether to factor out the GCF or its opposite from the second group to make them match; if you don't, you won't be able to finish the problem.

You've Got Problems

Problem 2: Factor the polynomial $12x^4 + 6x^3 + 14x + 7$.

Special Factoring Patterns

Sometimes, you hardly have to do any work at all to factor a polynomial. This appeals to me, because deep down, I am a very lazy man, and if I didn't have to feed myself or my family, I'd be happily working a meaningless job and living in squalor (as I proved beyond any reasonable doubt when I was a bachelor). In some circumstances, all you have to do is recognize that the polynomial at hand follows a certain pattern. To spot these handy, time-saving patterns, you'll need to be able to spot perfect squares and perfect cubes.

A *perfect square* is what you get when you multiply something by itself (in other words, the result of squaring something). For instance, $36w^4$ is a perfect square, since it's the result of something multiplied by itself: $(6w^2)(6w^2) = 36w^4$.

A *perfect cube* can be created by multiplying something by itself two times (in other words, the result of cubing something). Therefore, $-8y^3$ is a perfect cube, since $(-2y)(-2y)(-2y) = -8y^3$.

Here are the special factor patterns you should be able to recognize. Memorize the formulas, because in some cases, it's very hard to generate them without wasting a lot of time.

- **The difference of perfect squares:** If two squares are subtracted, $(a^2 - b^2)$, you can automatically rewrite the difference as a binomial product $(a + b)(a - b)$. For example, the polynomial $x^2 - 16$ is a difference of perfect squares, since $x^2 - 16 = (x)^2 - (4)^2$. (If you compare $x^2 - 16$ to the formula $a^2 - b^2$, $a = x$ and $b = 4$, as they are the numbers that generate the perfect squares.) Therefore, it can be factored as $(x + 4)(x - 4)$.
- **The difference of perfect cubes:** If two perfect cubes are subtracted, they can be factored as the product of a binomial and a trinomial: $(a^3 - b^3) = (a - b)(a^2 + ab + b^2)$. For example, given the polynomial $8x^3 - 27$ (in this case, $a = 2x$ and $b = 3$), its factored form is $(2x - 3)((2x)^2 + (2x)(3) + (3)^2)$, or $(2x - 3)(4x^2 + 6x + 9)$.
- **The sum of perfect cubes:** Unlike the sum of perfect squares, when perfect cubes are added together, they do follow a specific factor pattern. This pattern differs only slightly from its sister formula, the difference of perfect cubes; in fact, only a few signs are different: $(a^3 + b^3) = (a + b)(a^2 - ab + b^2)$. Consider the polynomial $y^3 = 64$; this represents the sum of perfect cubes $a^3 + b^3$ when $a = y$ and $b = 4$, so according to the formula, its factored form is $(y + 4)(y^2 - 4y + 16)$.

Talk the Talk

A **perfect square** is a quantity that results when something is multiplied by itself, and a **perfect cube** is the result of multiplying something by itself twice.

Kelley's Cautions

Note that you cannot factor the *sum* of perfect squares as easily as the *difference*. In fact, as far as you're concerned, you cannot factor $a^2 + b^2$, even though most algebra students try to factor it as $(a + b)(a + b)$, which is totally wrong! If you multiply $(a + b)(a + b)$, you get $a^2 + 2ab + b^2$, not $a^2 + b^2$. (If you need to review polynomial multiplication, it's covered in Chapter 10.)

One word of caution: Factoring by means of these special patterns works hand in hand with, not in opposition to, factoring using the greatest common factor. In fact, you should *always* look for a greatest common factor (and if it exists, factor it out) before trying to apply any other factoring technique. This guarantees that you'll get the fully factored form of the polynomial, and in some cases, will change the problem from impossible to easy.

Example 3: Factor the polynomials.

(a) $16x^3 + 2y^3$

Solution: This feels like a sum of perfect cubes problem, thanks to the powers of 3, but it doesn't exactly fit the formula. Even though x^3 is a perfect cube, $16x^3$ isn't. (There's no rational number that multiplied by itself twice gives you 16.) Same goes for the other term.

This is one example why you should always try to factor out the greatest common factor first. Notice that both terms have a GCF of 2, so factor it out.

$$2(8x^3 + y^3)$$

Ignore that 2 outside the parentheses for now; the quantity inside *is* the sum of perfect cubes, and fits the formula if $a = 2x$ and $b = y$. Rewrite it in factored form, leaving the GCF exactly where it is, out front.

$$2(2x + y)(4x^2 - 2xy + y^2)$$

CAUTION

Kelley's Cautions

Always assume that you're expected to factor a polynomial *completely*. In other words, none of the resulting factors should, in turn, be factorable.

(b) $x^4 - 16$

Solution: This is the difference of perfect squares, and fits the formula $a^2 - b^2 = (a + b)(a - b)$ if you set $a = x^2$ and $b = 4$. So, rewrite in the factored form prescribed by that formula:

$$(x^2 + 4)(x^2 - 4)$$

If you were to end here, you'd technically get the problem wrong, because it's not factored *completely;* one of the factors, $x^2 - 4$, is itself a perfect square and must be factored further.

$$(x^2 + 4)(x + 2)(x - 2)$$

You've Got Problems

Problem 3: Factor the polynomial $5x^2 - 125$.

Factoring Trinomials Using Their Coefficients

Now that you've learned how to factor out greatest common factors and can recognize important factor patterns, it's time to introduce the third most common factoring skill you'll need in algebra: factoring trinomials whose terms contain consecutive powers. (In other words, the exponents of the variables come one right after the other, like 0, 1, 2 or 3, 4, 5.) I'll spend the rest of this chapter dealing with this topic.

In this section, I'll describe a technique that does a great job of factoring trinomials whose leading coefficient is 1. In the final section of the chapter, I'll tell you how to handle polynomials with different leading coefficients.

The vast majority of the trinomials you'll factor are quadratic. In other words, you'll factor something that looks like this:

$$x^2 + ax + b, \text{ where } a \text{ and } b \text{ are integers}$$

(If you're wondering why there's no variable coefficient on the x^2 term, remember that the x^2's coefficient is also the leading coefficient for the polynomial, and it needs to be 1 for this technique to work properly.)

Your goal will be to factor that trinomial into two binomials that look like this:

$$(x + \boxed{?})(x + \boxed{?})$$

So, all you have to do is figure out what numbers go in those boxes. Here's the trick. If the trinomial $x^2 + ax + b$ is factorable, then two numbers will exist whose *sum* is a (the coefficient of the x term) and whose product is b (the constant). All you have to do is experiment until you figure those numbers out.

Once you know what they are, plug each one into its own box (it doesn't matter which one), and you're done!

Critical Point

When you're trying to figure out what goes into the boxes to factor $x^2 + ax + b$, pay special attention to the signs of a and b. If b is positive, then the mystery numbers are either both positive or both negative. If b is negative, then one of the mystery numbers must be positive and the other must be negative.

Example 4: Factor the polynomials.

(a) $x^2 + 7x + 12$

Solution: Ask yourself, "What two numbers add up to 7 but multiply to give you 12?" Since 12 is positive, both mystery numbers must be positive or negative, as I explained in a nearby margin note. Furthermore, since the sum of the mystery numbers, a, is also positive, then the only logical conclusion is that the numbers you seek *must* both be positive. (There's no way to add up two negative numbers to get a positive sum.)

It helps to follow a pattern if the answer doesn't dawn on you right away. Since you know the sum is 7, try two basic positive numbers whose sum is 7: 6 + 1. However, $6 \cdot 1 \neq 12$, so that's not right. Okay, so try 5 + 2 next; those numbers also have a sum of 7. Unfortunately, $5 \cdot 2 = 10$, which is not right, but it's a lot closer to the product you're looking for.

Next try 4 + 3. Hey … this looks good. The sum is 7 and $4 \cdot 3 = 12$. Bingo! All you have to do is plug those numbers into $(x + \boxed{?})(x + \boxed{?})$ and you're done: $(x + 3)(x + 4)$. Since multiplication is commutative, you could also have given an answer of $(x + 4)(x + 3)$ and been equally correct.

(b) $w^2 - 3w - 54$

Solution: Even though you've got w's instead of x's, the process is the same. Your answer will just have w's instead of x's. Since the constant, –54, is negative, you know one of the mystery numbers is positive and one is negative. In addition, they should add up to –3, so you can also surmise that the negative number is bigger than the positive number.

If you can't figure out where to start, try the procession of terms with sum –3: –4 + 1, –5 + 2, –6 + 3, etc. You'll eventually get the numbers –9 and 6, which have the correct sum and product, so the factored form of the polynomial is $(x - 9)(x + 6)$.

Critical Point

Another way to find the mystery numbers in Example 4(b) is to start with the constant, rather than the x coefficient. Instead of trying pairs of numbers that add up to –3, try pairs of numbers whose product is –54.

You've Got Problems

Problem 4: Factor the polynomial $2x^3 - 24x^2 + 64x$. *Hint: Once you factor out the greatest common factor, it will look like the problems in Example 4.*

Factoring with the Bomb Method

I'll be honest; no one but me calls this little factoring technique "the bomb." In fact, I know of few people that teach it at all, because when it comes to factoring tricky quadratic trinomials (tricky because their leading coefficients aren't 1), most algebra teachers tell you to "play around" with binomials until you find something that works.

That is a startling lack of direction in the face of truly tough factoring problems, especially considering that just about everything else you learn in algebra has 1,241,933 steps you have to follow in *exactly* the right order, or else the ghosts of long-dead mathematicians rise from their graves, kick you in the shins, and mark points off your test.

I learned this method many moons ago, under the name "factoring by decomposition," but that name sounds more like a term tossed around at a mortuary (speaking of dead mathematicians). It deserves a name that says "I am an all-but-forgotten method that will help you factor any *nonprime* trinomial in the known universe," so I decided to rename it.

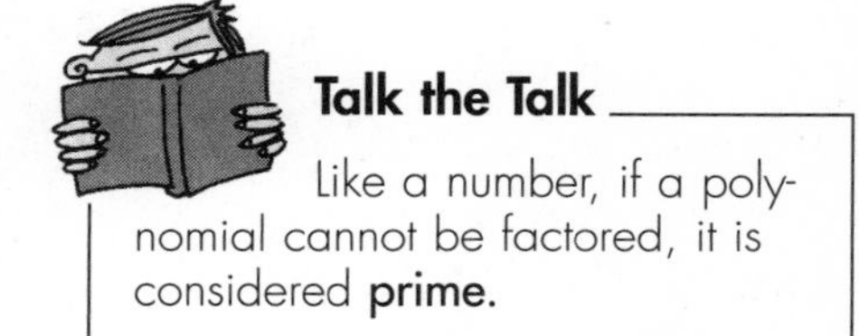

Talk the Talk

Like a number, if a polynomial cannot be factored, it is considered **prime**.

Here are the steps that comprise the bomb technique of factoring the polynomial $ax^2 + bx + c$:

1. **Factor out the GCF, if it exists.** This should always be your first step in every factoring problem.
2. **Find two mystery numbers.** These are similar to the numbers you sought when factoring simpler trinomials, but just a bit different. You still want the sum to be the x-coefficient (b), but now you want their product to equal ac, the product of the leading coefficient and the constant.
3. **Replace the x-coefficient.** Rewrite the polynomial, but where b once stood, write the sum of the two mystery numbers in parentheses. This powerful little clump of parentheses is about to blow this trinomial up into two binomials; that's why I call this method the bomb.
4. **Distribute x through the parentheses.** There's still an x multiplied times the quantity in parentheses from step 3; multiply each of the mystery numbers by x to eliminate those parentheses.
5. **Factor by grouping.** The end result, as it was before when you factored by grouping, will be the product of two binomials.

The best way to get really good at this method is to practice a lot. Make up your own practice trinomials to factor by multiplying simple binomials together; and then try to factor the result back into those binomials.

Example 5: Factor the polynomial $6x^2 - x - 12$.

Solution: This polynomial has no GCF, so you skip right to calculating the mystery numbers; they should add up to –1 and have a product of $6(-12) = -72$. Those numbers, then, must be –9 and 8. Replace the x coefficient of –1 with the sum of those mystery numbers in parentheses.

$$6x^2 + (-9 + 8)x - 12$$

Distribute the x lying outside the parentheses.

$$6x^2 - 9x + 8x - 12$$

Now you can factor by grouping to finish.

$$3x(2x - 3) + 4(2x - 3)$$

$$=(2x - 3)(3x + 4)$$

You've Got Problems

Problem 5: Factor the polynomial $4x^2 + 23x - 6$.

The Least You Need to Know

- Start every factoring problem by checking for a greatest common factor.
- Differences of perfect squares can be factored, as can both sums and differences of perfect cubes. However, sums of perfect squares have no common factor pattern.
- When factoring a quadratic polynomial whose leading coefficient is 1, just find two numbers that add up to the x coefficient and whose product is the constant.
- In the bomb method of factoring trinomials of type $ax^2 + bx + c$, you replace b with the sum of two mystery numbers that add up to b and multiply to give you ac.

Chapter 12

Wrestling with Radicals

In This Chapter

- Simplifying radical expressions
- Performing operations on radical expressions
- Solving simple equations containing radicals
- Dealing with imaginary numbers

Exponential powers are old news. You've squared things, cubed things, occasionally raised things to a negative power, and even gotten into a shoving match with a variable raised to the fifth power once, outside a shady nightclub on the edge of town. (If his coefficient hadn't been there, who knows what would have happened?)

Let's be frank. Exponents have been strutting around and showing off. They're at the top of the food chain, with no natural enemies. Sure, addition's great, but it has to look out for its arch nemesis, subtraction, and multiplication has never been the same since division showed up and started talking to his girlfriend. However, the worm has turned, and exponents are about to be put in their respective places; you're now going to learn about radicals, little symbols that keep exponential powers in check (which is quite a coincidence, since the radical symbol almost looks like a little check mark).

Introducing the Radical Sign

A *radical expression* looks like this: $\sqrt[a]{b}$ (read "the *a*th root of *b*"), and is made up of three major parts:

> **Talk the Talk**
>
> The **radical expression** $\sqrt[a]{b}$ has three major features, the **radical symbol** (it looks like a check mark), the **index** (the small number tucked outside the radical symbol), and the **radicand,** the quantity written beneath the horizontal bar of the radical symbol.

- *Radical symbol:* the symbol that looks like a check mark with an elongated horizontal line stuck to the end of it.
- *Index:* the small number tucked inside the check mark portion of the radical sign; in the expression $\sqrt[a]{b}$, *a* is the index.
- *Radicand:* the quantity written inside the radical symbol, beneath its horizontal roof; *b* is the radicand of the radical expression $\sqrt[a]{b}$.

Most of the radicals you'll deal with will have an index of 2, and are called *square roots.* In fact, they're so common that if no explicit index is written in a radical expression ($\sqrt{13}$ for example), you automatically assume that the index is 2.

> **Talk the Talk**
>
> Radicals with an index of 2 are called **square roots;** radicals with an index of 3 are called **cube roots.** While an index can be any natural number, only these two kinds of radicals have special names.

Simplifying Radical Expressions

Think of a radical symbol like a prison, and the pieces of the radicand as inmates. Not all the prisoners are doomed to a life sentence, trapped inside the dank (and foul-smelling) radical big house—there is a chance of parole. However, in order to be released from the radical sign, you must meet its parole requirements.

Specifically, a radical will only release things raised to a power that matches its index. So, square roots will only release pieces of its radicand that are raised to the second power, and a radical with an index of 5 will only release things raised to the fifth power. When asked to simplify radicals, what you're actually doing is paroling the factors within that meet the requirements, and leaving the rest inside to rot.

Example 1: Simplify the radical expressions.

(a) $\sqrt[3]{16x^4y^6}$

Solution: Start by factoring the radicand's coefficient; in other words, write it as a product of smaller numbers. However, since the index of the radical is 3, you want the factors to be powers of 3 if possible. Therefore, instead of writing 16 as $16 \cdot 1$ or $4 \cdot 4$, write it as $8 \cdot 2$, or $2^3 \cdot 2$, since 8 is a perfect cube and can be written as 2^3.

$$\sqrt[3]{2^3 \cdot 2 \cdot x^4 y^6}$$

Now turn your attention to the variables. You can rewrite x^4 as $x^3 \cdot x$ (since $x^3 \cdot x = x^{3+1} = x^4$), so it contains an exponent of 3. Luckily, y^6 is a perfect cube, since $y^2 \cdot y^2 \cdot y^2 = y^6$, so write it as with that all-important power of 3 as well: $(y^2)^3$.

$$\sqrt[3]{2^3 \cdot 2 \cdot x^3 \cdot x \cdot \left(y^2\right)^3}$$

Of all the pieces in the radicand, only 2^3, x^3, and $(y^2)^3$ contain powers of 3. Yank them out in front of the radical, stripping away the third power as they exit the prison, which leaves only 2 and x inside.

$$\left(2xy^2\right)\sqrt[3]{2x}$$

Critical Point

To make simplifying radicals easier, memorize the first 15 perfect squares (1, 4, 9, 16, 25, 36, 49, 64, 81, 100, 121, 144, 169, 196, 225) and the first 5 perfect cubes (1, 8, 27, 64, 125).

Critical Point

Why do you rewrite y^6 as $(y^2)^3$ in Example 1(a)? Basically, you're trying to make groups of three things, so that they can be released from the radical. It might help to think of $(y^2)^3$ as a group of three y^2's, and $(y^2)^3 = y^6$ thanks to exponential Rule 3 from Chapter 3.

(b) $\sqrt{18x^2y^3}$

Solution: Since this is a square root, you want as much of the radicand as possible to be raised to the second power. The coefficient 18 only has one factor that's a perfect square: 9; so, rewrite 18 as the product $2 \cdot 9$ (or $2 \cdot 3^2$). The x^2 term already has an exponent of 2, but you should rewrite the y^3 term as $y^2 \cdot y$, to identify the y^2 as a candidate for release.

$$\sqrt{2 \cdot 3^2 \cdot x^2 \cdot y^2 \cdot y}$$

Pull everything with an exponent of 2 outside the radical (and take away the power as you do), leaving what's left as the new radicand.

$$3|xy|\sqrt{2y}$$

You've got to be wondering where the heck those absolute value signs came from! I kind of sprung them on you, and I apologize. However, there's a rule in algebra that says if you ever have the expression $\sqrt[n]{x^n}$ (the power of the exponent matches the index of the radicand), and n is even, then the simplified result is $|x|$. (This is because you always want an even-powered root to have a positive answer, and those absolute values make sure that no matter what the variable's value is, the answer will be positive.)

CAUTION

Kelley's Cautions

Don't forget, if you have a variable in a radicand raised to the same, even power as its index, you should surround it with absolute value symbols once it is released.

Since you're releasing both x^2 and y^2 from this square root, and those exponents match the index of the radical, you've got to toss them inside absolute value signs once they're paroled.

You've Got Problems

Problem 1: Simplify the radical $\sqrt{300x^6y^3}$.

Unleashing Radical Powers

Radicals can actually be expressed without a radical sign by means of fractional exponents. The expression $a^{1/2}$ (read "a to the one-half power") is equivalent to $\sqrt{a}$, and $b^{1/3}$ means the same thing as $\sqrt[3]{b}$. However, the fractional exponent does not always have a 1 as its numerator; you can use other numbers to create more complicated radical expressions.

Critical Point

Remember, the denominator of a fractional power is the index of the equivalent radical, and the numerator is a power, either of the *radicand* or of the entire radical.

In general, the expression $x^{a/b}$ is equivalent to both $\sqrt[b]{x^a}$ and $\left(\sqrt[b]{x}\right)^a$; even though they look different, both mean the exact same thing, so it doesn't matter which one you choose. However, you'll usually find that the first one is more useful when simplifying variable expressions and the second is much handier for numbers.

Example 2: Simplify the expressions.

(a) $64^{2/3}$

Solution: Rewrite $64^{2/3}$ as $\left(\sqrt[3]{64}\right)^2$. The order of operations tells you to simplify parentheses first, and since the radical has an index of 3, you're looking for perfect cubes. Luckily, $64 = 4^3$, and therefore $\left(\sqrt[3]{64}\right)^2 = \left(\sqrt[3]{4^3}\right)^2$. Since 4 has a power of 3 within a radical of index 3, it is paroled, and you'll get $4^2 = 16$. So, $64^{2/3} = 16$. If you had rewritten $64^{2/3}$ as $\sqrt[3]{64^2} = \sqrt[3]{4096}$, the answer is still 16; it's just harder to see that 4096 is a perfect cube ($16^3 = 4096$).

(b) $4^{5/3}$

Solution: Rewrite the expression as $\left(\sqrt{4}\right)^5$. Happily, 4 is a perfect square: $2^2 = 4$.

$$\left(\sqrt{4}\right)^5 = \left(\sqrt{2^2}\right)^5 = 2^5 = 32$$

You've Got Problems

Problem 2: Simplify the expression $25^{3/2}$.

Radical Operations

When I say "radical operations," I don't mean extreme medical procedures, like having your arm removed and replaced with an otter, or perhaps having your cousin Irving surgically grafted to your left side, so that you become the world's first man-made conjoined twins. Instead, I mean the much more boring concepts of adding, subtracting, multiplying, and dividing radicals.

There are specific rules you have to follow when simplifying expressions containing radical expressions, just like there are special rules governing polynomial expressions. Once again, you'll find that multiplication and division don't have the same strenuous requirements that addition and subtraction have placed on them.

Addition and Subtraction

Back in Chapter 10, you learned that you could only add or subtract two polynomial terms together if they had the exact same variables; terms with matching variables were called "like terms." Radicals operate in a very similar way. In order to add two radicals together, they must be *like radicals;* in other words, they must contain the *exact* same radicand and index.

If you're asked to add or subtract radicals that contain different radicands, don't panic. Try to simplify the radicals—that usually does the trick.

Example 3: Simplify the expression $3\sqrt{2xy}-2\sqrt{50xy}$.

Solution: Notice that the second radical can be simplified, since one of 50's factors is a perfect square. Rewrite $2\sqrt{50xy}$ as $\sqrt{5^2\cdot 2xy}$ and pull the 5 out of the radical to get $5\sqrt{2xy}$.

The original problem now looks like this:

$$3\sqrt{2xy}-2\left(5\sqrt{2xy}\right)$$

Simplify the second radical further by multiplying the 2 outside the parentheses by the coefficient within.

$$3\sqrt{2xy}-10\sqrt{2xy}$$

Without even trying to make it happen, you've created a pair of like radicals, since the radicands and indices match. Because they're like radicals, you can combine their coefficients (3 – 10 = –7) and follow the result with that matching radical expression. So, the final answer will be $-7\sqrt{2xy}$.

Talk the Talk

Like radicals have matching radicands and indices, like $6\sqrt[5]{2x^2y}$ and $-9\sqrt[5]{2x^2y}$; only like radicals can be added to or subtracted from one another.

You've Got Problems

Problem 3: Simplify the expression $\sqrt[3]{8x^4}+4x\sqrt[3]{x}$.

Multiplication

Radicals have one important property that I have not yet mentioned: If two radicals with the same index are multiplied together, the result is just the product of the radicands beneath a single radical of that index. Translation: If you're multiplying radicals with matching indices, just multiply what's underneath the radical signs together, and write the result under a radical sign with the same index as the original radicals had.

$$\left(\sqrt[n]{x}\right)\left(\sqrt[n]{y}\right)=\sqrt[n]{xy}$$

Notice that you don't need like terms in order to multiply radicals; all you need is that matching index.

Example 4: Simplify the expression $\left(\sqrt[3]{9x^4y^5}\right)^2$.

Solution: You're asked to square that radical, which means it's multiplied by itself.

$$\left(\sqrt[3]{9x^4y^5}\right)\left(\sqrt[3]{9x^4y^5}\right)$$

Multiply the radicands together and write the product beneath a radical sign of index 3:

$$\sqrt[3]{9\cdot 9\cdot x^4\cdot x^4\cdot y^5\cdot y^5}=\sqrt[3]{81x^8y^{10}}$$

Now all you need to do is simplify the radical.

$$\sqrt[3]{3^3\cdot 3\cdot\left(x^2\right)^3\cdot x^2\cdot\left(y^3\right)^3\cdot y}=\left(3x^2y^3\right)\sqrt[3]{3x^2y}$$

How'd You Do That?

This property of radicals is true thanks to exponential Rule 4 from Chapter 3, which said that $(xy)^a = x^ay^a$. Technically speaking, a radical is the same as a fractional power, so $\sqrt[n]{xy}$ is the same as $(xy)^{1/n}$, which can be rewritten as $x^{1/n}y^{1/n}$, or $\left(\sqrt[n]{x}\right)\left(\sqrt[n]{y}\right)$.

You've Got Problems

Problem 4: Simplify the product $\left(\sqrt{12x^2y}\right)\left(\sqrt{3xy}\right)$.

Division

The quotient of two radicals with the same index can be rewritten beneath a single radical sign, just like the product of those radicals. In other words, the expression $\sqrt{x}\div\sqrt{y}$ is equivalent to $\sqrt{\frac{x}{y}}$. However, there is a new concern that surfaces when you're dealing with radical division—the presence of a radical in the denominator of your final answer.

If an expression contains a radical symbol that cannot be completely eliminated during simplification, then it is most likely irrational (check back to Chapter 1 if you're not sure what that means). For a long time, it has been considered bad math etiquette to leave a radical in your answers, because of the long and ugly decimals associated with such numbers. It's the equivalent of going to a fancy dinner party and chewing with your mouth open, which would be considered rude, even if you don't mean it to be.

It's not as though you're actually expected to divide that horrible decimal into the numerator or anything, but even the *prospect* of such a gross division problem has created peer pressure to eliminate any radicals housed below the fraction line in a process called *rationalizing the denominator*. It's a final, and easy, step some teachers require and others (such as myself) don't. Make sure to ask whether you'll be expected to rationalize denominators in solutions.

Example 5: Calculate and rationalize the quotient.

$$\sqrt{12x^7y} \div \sqrt{8xy^3}$$

Solution: Write the quotient as a fraction beneath a single radical sign.

$$\sqrt{\frac{12x^7y}{8xy^3}}$$

You can simplify the coefficients by dividing both 12 and 8 by 4, the greatest common factor. Apply exponential Rule 2 from Chapter 3 to simplify the variables: $x^{7-1} = x^6$ and $y^{1-3} = y^{-2}$ (which means there's a y^2 in the denominator).

$$\sqrt{\frac{3x^6}{2y^2}}$$

Talk the Talk

The process of removing all radical quantities from the denominator of a fraction is called **rationalizing the denominator.**

Critical Point

Remember, you're allowed to multiply a fraction by anything divided by itself, because that's technically the same thing as multiplying by 1 (anything divided by itself equals 1).

Write the numerator and denominator as separate radicals again, and simplify them.

$$\frac{\sqrt{3x^6}}{\sqrt{2y^2}} = \frac{x^3\sqrt{3}}{|y|\sqrt{2}}$$

This fraction has an irrational piece in its denominator: $\sqrt{2}$. To eliminate it, multiply that value times both the numerator and denominator.

$$\frac{x^3\sqrt{3}}{|y|\sqrt{2}} \cdot \frac{\sqrt{2}}{\sqrt{2}}$$

Multiply the numerators and denominators separately and simplify. Notice the answer has no radical in the denominator.

$$\frac{x^3\left(\sqrt{6}\right)}{|y|\sqrt{4}} = \frac{x^3\sqrt{6}}{2|y|}$$

You've Got Problems

Problem 5: Find the quotient and rationalize, if necessary.

$$\sqrt{2x^2y^3} \div \sqrt{18x^3y^2}$$

Solving Radical Equations

Did you ever read *Alice in Wonderland?* It had a (slightly less popular) sequel called *Through the Looking Glass*, in which Alice actually passes into a mirror, and the world she encounters is a weird reflection of her own. Some physicists have likened the relationship between these worlds to the relationship between matter and antimatter.

Without getting too nerdy or discussing *Star Trek* propulsion systems (which are theoretically powered by matter/antimatter engines), I'll suffice it to say that you should never have matter and antimatter at the same social gathering, because if they come in contact with one another, there'll be a cataclysmic explosion. The two kinds of matter are the exact antitheses of one another, so much so that when they touch, they cancel one another out in the most permanent way possible: Detonation.

A slightly less violent, but similar, relationship exists between me and the girl I dated in high school, but that's irrelevant. More important for our purposes, the same relationship exists between radicals with index n and exponential expressions with power n. If they meet, they'll say, "It wasn't me that changed, it was you" (actually, that sounds more like my ex-girlfriend) and promptly explode, leaving behind only the contents of the radicand, a pile of smoke, and some embarrassing old love letters you can't remember writing.

This explosive property can be used both at heavy metal concerts to power the pyrotechnics of groups like Limozeen and Taranchula, or to solve equations containing radicals. In the interest of space, I'll focus on the latter.

Talk the Talk

My heavy metal band names are a tribute to Strong Bad, a cartoon character at the website www.homestarrunner.com. If you haven't happened upon this website yet, make sure you pay them a visit, and tell Marzipan I said hi.

Example 6: Solve the equation $\sqrt[3]{2x-1}+3=6$.

Solution: Since the variable is trapped inside that radical sign, you'll need to free it with some specially designed explosives. Before you set the charge, get everything but the radical away for safety. In other words, isolate the radical by subtracting 3 from both sides of the equation.

$$\sqrt[3]{2x-1}=3$$

To destroy a radical with index 3, you'll raise it to the third power. (If it had been index 5, you'd raise both sides to the fifth power to eliminate it; just match the exponent with the index.) To keep the equation balanced, you should raise its other side to the third power as well.

$$\left(\sqrt[3]{2x-1}\right)^3 = 3^3$$

All that remains on the left side of the equation are the smoldering contents of the radicand; the resulting equation is very simple to solve.

$$\begin{aligned} 2x - 1 &= 27 \\ 2x &= 28 \\ x &= 14 \end{aligned}$$

If you'd like, check your answer by substituting 14 for x in the original problem; you'll get $\sqrt[3]{27} + 3 = 6$, which is a true statement, which means you got the solution correct.

You've Got Problems

Problem 6: Solve the equation $2\sqrt{x-3} = 8$.

When Things Get Complex

So far in this chapter, the vast majority of the radicands you've seen have been positive. They can't help it—they're just upbeat people, and there's nothing wrong with that. However, you do need to know how to handle it when a bit of negativity creeps in.

Sometimes a negative is no problem. Specifically, if the index of a radical is odd, then a negative radicand is completely valid. For instance, $\sqrt[3]{-8x^3}$ can be simplified as $-2x$, since $(-2x)(-2x)(-2x) = -8x^3$. Basically, a negative thing multiplied by itself an odd number of times will also be negative. However, if a radical has an even index, there's trouble.

While the expression $\sqrt{16}$ is easy to simplify ($\sqrt{16} = 4$, since $4 \cdot 4 = 16$), the radical expression $\sqrt{-16}$ is not. What times itself can equal a negative number? If you remember back to Chapter 1, the only way to multiply two numbers together and get a negative was when the two numbers had different signs, and there's no way something can have a different sign than itself!

There's Something in Your *i*

Luckily, math people are resilient folks, and they can invent stuff that allows even the most impossible things to happen. One of their handiest inventions is *i*, a letter representing the solution to a problematic negative radicand. The letter *i* is short for "imaginary number," and has the value $i = \sqrt{-1}$. If you think about it, *i* would have to be imaginary, because no real number could have the value $\sqrt{-1}$, for all the reasons I laid out above.

A number containing *i*, such as $2i$ or $-5i$, is said to be an *imaginary number*. Furthermore, any number of the form $a + bi$ (where a and b are real numbers) is said to be *complex*. Basically, a complex number is made up of an imaginary part, bi, added to or subtracted from a real part, a. For example, in the complex number $4 - 7i$, the real part is 4 and the imaginary part is $-7i$.

Every complex number $a + bi$ has a *conjugate* paired with it, which is equal to $a - bi$; in other words, the only difference between a complex number and its conjugate is the sign preceding the imaginary part. For the complex number example I used a few moments ago, $4 - 7i$, the conjugate would be $4 + 7i$.

Probably the most important thing to remember about complex and imaginary numbers is that $i^2 = -1$, since $i^2 = \left(\sqrt{-1}\right)^2$; remember, you just learned that if a radicand is raised to an exponent that matches its index, both disappear, leaving behind only what was beneath the radical (–1 in this case). I know; it's weird that a number squared could be negative, but it's only true for imaginary numbers.

Talk the Talk

An **imaginary number,** bi, is the product of a real number b and the imaginary piece $i = \sqrt{-1}$. A **complex number** has the form $a + bi$, where a and b are both real numbers; every complex number is paired with a **conjugate,** $a - bi$, which matches the complex number exactly, except for the sign preceding bi.

How'd You Do That?

Every imaginary number is automatically a complex number as well. For instance, $3i$ has form $a + bi$ (making it complex) if $a = 0$ and $b = 3$. Additionally, every real number is automatically complex as well. Consider the real number 12; it has complex form $a + bi$ if $a = 12$ and $b = 0$.

Example 7: Simplify the expressions.

(a) 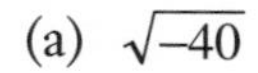 $\sqrt{-40}$

Solution: Rewrite $\sqrt{-40}$ as $\sqrt{-1 \cdot 4 \cdot 10}$. The perfect square ($4 = 2^2$) can be pulled out of the radical, and $\sqrt{-1}$ can be rewritten as *i*: $2i\sqrt{10}$.

(b) i^5

Solution: Since you know $i^2 = -1$, rewrite i^5 using as many i^2 factors as possible.

$$i^5 = i^2 \cdot i^2 \cdot i$$

(The sum of the exponents on the right side, 2 + 2 + 1, must equal 5, the exponent on the left side.) Now replace each i^2 with –1.

$$i^5 = (-1)(-1)i$$

$$i^5 = i$$

You've Got Problems

Problem 7: Simplify the expression $\sqrt{-36}+i^3$.

Simplifying Complex Expressions

You'll need to know how to add, subtract, multiply, and divide complex numbers, but every complex number is really just a binomial, so you'll apply the same methods in Chapter 10 that you used with polynomials (except when it comes to division, that is). Here's a quick rundown describing how the four major operations work with complex numbers:

- **Addition:** Since imaginary numbers contain the same variable, *i*, treat them as like terms. Just add up the real parts and the imaginary parts separately.

$$(3 - 4i) + (2 + 9i) = 3 + 2 - 4i + 9i = 5 + 5i$$

- **Subtraction:** Distribute the negative sign, and all that's left behind is a simple addition problem.

$$(3 - 4i) - (2 + 9i) = 3 - 4i - 2 - 9i = 1 - 13i$$

- **Multiplication:** Just like any product of binomials, distribute each term in the first complex number through the second complex number separately.

$$(2 + 3i)(5 - 2i) = 2 \cdot 5 + 2(-2i) + 3i \cdot 5 + 3i(-2i)$$
$$= 10 - 4i + 15i - 6i^2$$

Replace i^2 with –1 and combine like terms.

$$10 - 4i + 15i - 6(-1)$$
$$= 10 - 4i + 15i + 6$$
$$= 16 + 11i$$

- **Division:** Good news! You don't have to do long or synthetic division to calculate the quotient of complex numbers. To calculate $(1 - i) \div (2 + 7i)$, start by writing the quotient as a fraction.

$$\frac{1-i}{2+7i}$$

Multiply both the numerator and denominator by the conjugate of the denominator.

$$\frac{1-i}{2+7i} \cdot \frac{2-7i}{2-7i}$$

Multiply the numerators together and write the result over the product of the denominators.

$$\frac{2-7i-2i+7i^2}{4-14i+14i-49i^2} = \frac{2-9i-7}{4+49} = \frac{-5-9i}{53}$$

If you want, you can write each term of the numerator divided separately by the denominator to get the result into the official $a + bi$ form of a complex number.

$$-\frac{5}{53} - \frac{9i}{53}$$

Remember, no simplified complex number will ever contain an i^2 in it; you should always replace that with –1 and simplify like terms.

You've Got Problems

Problem 8: Given the complex numbers $c = 3 - 4i$ and $d = 8 + i$, calculate (a) $c + d$, (b) $c - d$, (c) $c \cdot d$, and (d) $c \div d$.

The Least You Need to Know

- To simplify radicals, look for things inside the radical whose exponent matches the radical's index.
- You can only add or subtract like radicals.
- The expression $x^{a/b}$ can be rewritten as either $\sqrt[b]{x^a}$ or $\left(\sqrt[b]{x}\right)^a$.
- The complex number $a + bi$ consists of the sum of a real number a and an imaginary number bi.
- The conjugate of the complex number $a + bi$ is $a - bi$.

Chapter 13

Quadratic Equations and Inequalities

In This Chapter

- Finding solutions by factoring
- Completing the square to form binomial perfect squares
- Applying the quadratic formula
- Solving and graphing simple quadratic inequalities

When I taught math at a public high school, I also coached track and field for three years. That may shock you (since I am obviously a math geek), and quite honestly, it should. I have never and will never be known for being nimble or possessing quick, dexterous reflexes. In fact, I actually got kicked out of a gymnastics program when I was in elementary school, because they thought I was going to *paralyze* myself. I am the exact opposite of a cat—during a fall, most cats right themselves midair and manage to land on their feet. In gymnastics, no matter what position I fell from, or how high I was from the ground, I always managed to fall right on the top of my head.

Regardless, the school administration saw fit to make me coach of the track team, and immediately assigned me the task of training the hurdlers. (As coach, I only tried to hurdle once, and I think I must have knocked myself out doing it, because when I regained consciousness, my face was planted firmly in the track. Interestingly enough, my feet were still propped up on the hurdle.) Even though I was not the most proficient coach in the world, I did learn one thing—always start training runners with the hurdles at their lowest possible setting. There's no sense starting a new athlete out with the hurdle at its actual, intimidating race height right away. You start easy, and then slowly build the difficulty along the way.

You already know how to solve linear equations, so now I'm going to move the hurdles up a notch. Instead of solving equations containing polynomials of degree 1 (like those equations you solved way back in Chapter 4), it's time to throw in some polynomials of degree 2. You'll have to use *completely* different techniques than you did for linear equations, but the good news is that you've got three different methods to choose from to do so. By the end of the chapter, you'll be leaping like a gazelle over the new, higher hurdle, and I'll be quite proud of you, even though I'm down here, unconscious on the ground.

Solving Quadratics by Factoring

If you can transform an equation into a factorable quadratic polynomial, it is very simple to solve. Even though this technique will not work for all quadratic equations, when it does, it is by far the quickest and simplest way to get an answer. Therefore, unless a problem specifically tells you to use another technique to solve a quadratic equation, you should try this one first. If you get a prime (unfactorable) polynomial, you can always shift to one of the other techniques you'll learn later in this chapter.

To solve a quadratic equation by factoring, follow these steps:

1. **Set the equation equal to 0.** Move *all* of the terms to the left side of the equation by adding or subtracting them, as appropriate, leaving only 0 on the right side of the equation.

2. **Factor the polynomial completely.** Use one of the techniques you learned in Chapter 11 to factor; remember to always factor out the greatest common factor first.

3. **Set each of the factors equal to 0.** You're basically creating a bunch of tiny, little equations whose left sides are the factors and whose right sides are each 0. It's good form to separate these little equations with the word "or," because any one of them could be true.

How'd You Do That?

If you have the equation $(x - a)(x - b) = 0$, Step 3 tells you to change that into the twin equations:

$$x - a = 0 \text{ or } x - b = 0$$

Are you wondering why that's allowed? It's thanks to something called the *zero product property.*

Think about it this way: If two things are multiplied together—in this case the quantities $(x - a)$ and $(x - b)$—and the result is 0, then at least one of those two quantities actually has to be equal to 0! There's no way to multiply two or more things to get 0 unless at least one of them equals 0.

4. **Solve the smaller equations and check your answers.** Each of the solutions to the tiny, little equations is also a solution to the original equation. However, to make sure they actually work, you should plug them back into that original equation to verify that you get true statements.

The hardest part of this technique is actually the factoring itself, and since that's not a new concept, this procedure is very simple and straightforward.

Example 1: Solve the equations, and give all possible solutions.

(a) $x^2 - 6x + 9 = 0$

Solution: Since this equation is already set equal to 0, start by factoring the left side.

$$(x - 3)(x - 3) = 0$$

Now set each factor equal to 0.

$$x - 3 = 0 \text{ or } x - 3 = 0$$

$$x = 3 \text{ or } x = 3$$

Well, since both factors were the same, both solutions ended up equal, so the equation $x^2 - 6x + 9 = 0$ only has one valid solution, $x = 3$. When you get an answer like this, which appears as a possible solution twice, it has a special name—it's called a *double root.*

Talk the Talk

A **double root** is a repeated solution for a polynomial equation; it's the result of a repeated factor in the polynomial.

Check to make sure that 3 is a valid answer by plugging it back into the original equation.

$$x^2 - 6x + 9 = 0$$
$$3^2 - 6(3) + 9 = 0$$
$$9 - 18 + 9 = 0$$
$$0 = 0$$

There's no doubt that 0 = 0 is a true statement, so you got the answer right.

(b) $3x^2 + 10x = -4x + 24$

Solution: Your first job is to set this equal to 0; to accomplish this, add $4x$ to and subtract 24 from both sides.

$$3x^2 + 14x - 24 = 0$$

Factor the trinomial using the bomb method discussed in Chapter 11. The two mystery numbers you're looking for are –4 and 18.

$$3x^2 + (-4 + 18)x - 24 = 0$$
$$3x^2 - 4x + 18x - 24 = 0$$
$$x(3x - 4) + 6(3x - 4) = 0$$
$$(3x - 4)(x + 6) = 0$$

Set each factor equal to 0 and solve.

$$3x - 4 = 0 \quad \text{or} \quad x + 6 = 0$$
$$x = \tfrac{4}{3} \quad \text{or} \quad x = -6$$

Both of these answers work when you check them.

You've Got Problems

Problem 1: Give all the solutions to the equation $4x^3 = 25x$.

Completing the Square

The next technique you can use to solve quadratic equations is called *completing the square;* it forces the equation, against its will, to contain a perfect square, which can then be easily eliminated.

Completing the square isn't exactly the easiest way to solve quadratic equations; its strength lies in the fact that the process is repetitive and predictable. It's like that guy you knew in high school whose parents made him take his cousin (who usually looks like a perfect square) to the prom. Sure it was awkward, and neither of them really had a lot of fun, but at least their parents knew exactly how the evening was going to end. (After all, this was his *cousin*, so there was no prospect of post-dance "extracurricular activities.")

Here's the best news yet: Completing the square *will always work*, unlike the factoring method, which, of course, requires that the trinomial be factorable.

However, you need to learn one thing before I can show you how to complete the square: how to eliminate exponents in equations.

Solving Basic Exponential Equations

Last chapter you learned how to solve radical equations. During that process, you found out raising both sides of an equation to the nth power cancels out a radical with index n. In other words, to solve the equation

$$\sqrt[4]{x-1}=3$$

you would have to raise both sides of the equation to the fourth power, leaving you with

$$\left(\sqrt[4]{x-1}\right)^4=3^4$$
$$x-1=81$$
$$x=82$$

This process also works in reverse. In other words, you can cancel out an exponent of n by taking the nth root of both sides of the equation. For example, to solve the equation

$$5x^3=80$$

you'd first isolate the quantity raised to the exponent (just like you did in radical equations).

$$\frac{5x^3}{5}=\frac{80}{5}$$
$$x^3=16$$

To cancel out the exponent of 3, leaving behind only what was once raised to the third power, take the third root of both sides of the equation.

$$\sqrt[3]{x^3}=\sqrt[3]{16}$$
$$x=2\sqrt[3]{2}$$

Critical Point

Remember, last chapter you learned that $\sqrt[n]{x^n} = |x|$ when n is an even number. The "±" rule discussed here takes care of that for you when dealing with equations.

There's only one major thing to keep in mind when canceling out that exponent: If you're canceling out an even exponent, you need to stick in a "±" sign on the left side of the equation when you jam those radical signs on.

For example, consider the equation $x^2 = 16$. You can solve for x by square rooting both sides of the equation.

$$\sqrt{x^2} = \pm\sqrt{16}$$
$$x = \pm 4$$

Bugging Out with Squares

Now it's finally time to date your cousin, er, complete the square. Because there are a lot of steps involved, I should explain how they work in the context of an example. Just a brief warning before I get started: This problem will contain a scatological reference and is hence rated PG by the National Advisory Council on Lowbrow Humor in Mathematics.

Example 2: Solve the equation $2x^2 - 16x + 10 = 0$.

Solution: Even though you can factor out a GCF of 2, the resulting equation, $2(x^2 - 8x + 5) = 0$, cannot be factored further. So in order to solve it, you have to resort to completing the square.

Step 1: Make sure the coefficient of x^2 is 1. If it's not, divide everything in the equation by that coefficient. In this problem, the coefficient of x^2 is 2, so divide everything by 2.

$$x^2 - 8x + 5 = 0$$

Kelley's Cautions

If you forget to make the coefficient of x^2 equal to 1, you'll get stuck later on, in Step 4.

Step 2: Move everything to the left side of the equation except the constant. You want that constant sitting alone on the right side of the equation. In this example, that means subtracting 5 from both sides.

$$x^2 - 8x = -5$$

Step 3: Inspect bug droppings. This is the key step for completing the square: You're going to add a mystery number to both sides of the equation. To find that number, take half of the x coefficient (half of –8 is –4) and square it ($(-4)^2 = 16$). So, in this case, the mystery number to be added to both sides of the equation is 16.

I don't know when, why, or how I came up with it, but for years I have used a little, three-segmented bug (pictured in Figure 13.1.) to help me remember this step. To this day, I still draw his body (minus the legs, eyes, and other time-consuming details) in the margins of a completing the square problem.

The bug ingests the coefficient of the x (or middle) term.

He eats only half of the coefficient (half of –8 is –4).

He leaves a piece of "square waste" (–4 squared is 16).

Figure 13.1

Don't turn on a bright light while working with the bug, or he may scuttle under the fridge.

Here's how my little invertebrate mnemonic friend works: The bug eats the x coefficient (place the x coefficient, including its sign, in the top segment); however, he's only a bug with a small appetite, so he is only able to eat half of that number (put half of the number located in the first segment, including its sign, in the second segment). Finally, once that number's eaten, he must, ahem, excrete some waste to complete the digestive process, and this bug only leaves *square* droppings (the square of the middle segment is written in the bug's rear end).

Whether you choose to use the bug to generate the mystery number or not, you should add it (it will always be positive since it is the result of a square) to both sides of the equation.

$$x^2 - 8x + 16 = -5 + 16$$

$$x^2 - 8x + 16 = 11$$

Talk the Talk

This procedure is called **completing the square** because, by adding 16 to both sides in Example 2, you're creating a trinomial that's a perfect square. Therefore, it can be written as $(x + a)^2$, where a is the number from the bug's midsection.

Step 4: Rewrite the left side of the equation as a perfect binomial square. When factored, the quadratic on the left side will look like this: $(x + a)^2$. Here's the cool part: The a in $(x + a)^2$ comes right from the stomach of the bug! Since the bug has –4 in its stomach for this problem, the factored form should look like

$$(x - 4)^2 = 11$$

Step 5: Eliminate the exponent. To counteract the square, take the square root of both sides; don't forget to stick a "±" sign on the right side and simplify that radical (if possible) when you do.

$$\sqrt{(x-4)^2} = \pm\sqrt{11}$$
$$x-4 = \pm\sqrt{11}$$

Step 6: Solve for *x*. In this problem, add 4 to both sides.

$$x = 4 \pm \sqrt{11}$$

So, the quadratic equation $2x^2 - 16x + 10 = 0$ has two solutions: $x = 4+\sqrt{11}$ and $x = 4-\sqrt{11}$; the "±" sign means that either sign will work for the expression.

You've Got Problems

Problem 2: Solve the equation by completing the square.

$$x^2 + 6x - 3 = 0$$

The Quadratic Formula

The final method you can use to solve quadratic equations is called the *quadratic formula*, and it, like the completing-the-square method, will also solve any quadratic equation. Even better, it's very easy to use.

All you have to do is set the quadratic equation equal to 0, and it will look something like this:

$ax^2 + bx + c = 0$, where a, b, and c are real numbers

To get the solutions to the equation, just plug the coefficients a, b, and c into the quadratic formula:

$$x = \frac{-b \pm \sqrt{b^2 - 4ac}}{2a}$$

I know that formula looks ugly at first, but you're going to have to memorize it. If you need help doing that, check out my web page www.calculus-help.com and click on the "Fun Stuff" section; I wrote a little song (parental warning: song contains comedic violence) to help you remember it.

Example 3: Solve the equation using the quadratic formula.

$$2x^2 - 5x = -1$$

Solution: To set the equation equal to 0, into the form $ax^2 + bx + c = 0$, add 1 to both sides.

$$2x^2 - 5x + 1 = 0$$

So, $a = 2$, $b = -5$, and $c = 1$. Plug these values into the quadratic formula.

$$x = \frac{-b \pm \sqrt{b^2 - 4ac}}{2a}$$

$$x = \frac{-(-5) \pm \sqrt{(-5)^2 - 4(2)(1)}}{2(2)}$$

$$x = \frac{5 \pm \sqrt{25 - 8}}{4}$$

$$x = \frac{5 \pm \sqrt{17}}{4}$$

How'd You Do That?

Wondering where the quadratic formula comes from? Is it, too, the product of insect waste? Actually, in a manner of speaking, it is. The quadratic formula is the solution to the equation $ax^2 + bx + c = 0$ when it's solved by completing the square.

The solutions to the quadratic equation $2x^2 - 5x = -1$ are $x = \frac{5 + \sqrt{17}}{4}$ or $x = \frac{5 - \sqrt{17}}{4}$.

You could also rewrite the solutions as $x = \frac{5}{4} + \frac{\sqrt{17}}{4}$ or $x = \frac{5}{4} - \frac{\sqrt{17}}{4}$, but since none of those fractions can be reduced (simplified), there's no need to.

You've Got Problems

Problem 3: Solve the equation from Problem 2 ($x^2 + 6x - 3 = 0$) again, this time using the quadratic formula; verify that you get the same answer as before.

All Signs Point to the Discriminant

Have you ever owned one of those Magic 8 Balls? They look like comically oversized pool balls, but have a flat window built into them, so that you can see what's inside—a 20-sided die floating in disgusting opaque blue goo. Supposedly, the billiard ball has prognostic powers; all you have to do is ask it a question, give it a shake, and slowly, mystically, like a petroleum-covered seal emerging from an oil spill, the die will rise to the little window and reveal the answer to your question.

Talk the Talk

The **discriminant** is the expression $b^2 - 4ac$, which is defined for any quadratic equation $ax^2 + bx + c = 0$. Based upon the sign of the expression, you can determine how many real number solutions the quadratic equation has.

The quadratic equation contains a Magic 8 Ball of sorts. The expression $b^2 - 4ac$ from beneath the radical sign is called the *discriminant*, and it can actually determine for you how many solutions a given quadratic equation has, if you don't feel like actually calculating them.

Considering that an unfactorable quadratic equation requires a lot of work to solve (tons of arithmetic abounds in the quadratic formula, and a whole bunch of steps are required in the completing the square method), it's often useful to gaze into the mystic beyond to make sure the equation even *has* any real number solutions before you spend any time actually trying to find them.

Here's how the discriminant works. Given a quadratic equation $ax^2 + bx + c = 0$, plug the coefficients into the expression $b^2 - 4ac$ to see what results:

- If you get a positive number, the quadratic will have two unique solutions.
- If you get 0, the quadratic will have exactly one solution, a double root.
- If you get a negative number, the quadratic will have no real solutions, just two imaginary ones. (In other words, solutions will contain the i you learned about in Chapter 12.)

The discriminant isn't magic. It just shows how important that radical is in the quadratic formula. If its radicand is 0, for example, then you'll get $\frac{-b \pm \sqrt{0}}{2a}$, or $-\frac{b}{2a}$, a single solution. If, however, $b^2 - 4ac$ is negative, then you'll have a negative inside a square root sign in the quadratic formula, meaning only imaginary solutions.

Example 4: Without calculating them, determine how many real solutions the equation $3x^2 - 2x = -1$ has.

Solution: Set the quadratic equation equal to 0 by adding 1 to both sides.

$$3x^2 - 2x + 1 = 0$$

Set $a = 3$, $b = -2$, and $c = 1$, and evaluate the discriminant.

$$b^2 - 4ac$$

$$= (-2)^2 - 4(3)(1)$$

$$= 4 - 12$$

$$= -8$$

Because the discriminant is negative, the quadratic equation has no real number solutions, only two imaginary ones.

You've Got Problems

Problem 4: Without calculating them, determine how many real solutions the equation $25x^2 - 40x + 16 = 0$ has.

Solving One-Variable Quadratic Inequalities

So far, this chapter has been chock-full of procedures that require precise step-by-step instructions, and the final topic of the chapter will be no different. In fact, there have been so many steps to follow in the last few topics, you probably feel like you're putting together Ikea furniture, but bear with me just a little longer and you'll know all you need to know about quadratics.

Remember back in Chapter 7 when you learned that the solution to an inequality is often expressed as a graph, because there are almost always an infinite number of possible solutions? This is still true if the inequality contains a quadratic polynomial, like $2x^2 + x - 2 < 0$.

Let me jog your memory again. What do you use to graph an inequality containing only one variable—a number line or a coordinate plane? If you answered coordinate plane, oooh, I'm sorry, you're wrong, but at least you'll walk away with some fine parting gifts and a home version of our game. The correct answer is a number line; the number of axes in the graphing system should match the number of unique letters in the equation or inequality. For example, since $2x^2 + x - 2 < 0$ contains only x (not x and y both), you should use a number line to graph it.

That's enough talk for now; let me explain the steps you'll follow to solve a quadratic inequality:

1. **Pretend the inequality symbol is an equal sign and solve the corresponding quadratic equation.** Those solutions are called the *critical numbers* of the inequality. You're not changing that inequality sign to an equal sign permanently—once you find the critical numbers, you can change it back.

Talk the Talk

The **critical numbers** are the values of x for which an inequality equals 0 or is undefined. They break the number line into segments, called **intervals.**

2. **Graph the solutions on a number line.** Just mark the critical numbers on the number line with either open or closed dots, depending upon whether or not the symbol allows for the possibility of equality (just like you did during the number line graphs of Chapter 7).
3. **Use test values to find the solution interval(s).** The critical numbers will split the number line into segments, called *intervals*. Choose a test value from each interval and plug it into the original inequality for x. If the test value makes the inequality true, then so will all of the other values in that same interval, so it is part of the solution.

Even though the graph you create during this process tells you what the solution is, most teachers would rather you not just give the graph as an answer. They usually want you to express that interval as an inequality statement.

Example 5: Solve the inequality and graph the solution.

$$2x^2 + x - 2 < 0$$

Solution: Pretend, for a moment, that this is actually the equation $2x^2 + x - 2 = 0$. Unfortunately, this quadratic cannot be factored, so you'll have to use either the quadratic formula or complete the square to get the solutions, which will be

$$x = -\frac{1}{4} + \frac{\sqrt{17}}{4} \text{ or } x = -\frac{1}{4} - \frac{\sqrt{17}}{4}$$

Boy, those critical numbers sure are ugly. It's a good idea to type them into a calculator to figure out what decimal they're approximately equal to, so that you can plot them on the number line as I have done in Figure 13.2. (Since the inequality symbol, <, does not allow equality, you have to use open dots.)

$$x = -\frac{1}{4} + \frac{\sqrt{17}}{4} \approx -0.25 + 1.03 \approx .78$$

$$x = -\frac{1}{4} - \frac{\sqrt{17}}{4} \approx -0.25 - 1.03 \approx -1.28$$

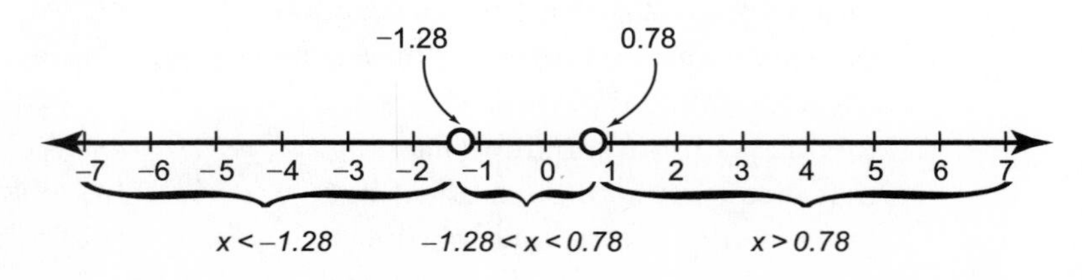

Figure 13.2

The critical numbers break the number line into the three labeled intervals. Notice that when both endpoints of an interval are critical numbers, it's useful to write that interval as a compound inequality.

Now choose one value to represent each interval (I chose the simplest ones I could: $x = -2$, $x = 0$, and $x = 1$) and plug them into the original inequality, to see which values result in true statements.

Test $x=-2$	Test $x=0$	Test $x=1$
$2(-2)^2+(-2)-2<0$	$2(0)^2+(0)-2<0$	$2(1)^2+(1)-2<0$
$2(4)-2-2<0$	$2(0)-2<0$	$2(1)+1-2<0$
$8-4<0$		$2-1<0$
$4<0$ False	$-2<0$ True	$1<0$ False

The interval containing $x = 0$ (the middle interval of Figure 13.2) is the only one that makes the inequality true, so it is the correct solution: $-1.28 < x < .78$. Of course, if you want to write the answer *exactly* right, you'd have to use the actual, ugly critical numbers, and not their decimal equivalents:

$$-\frac{1}{4}-\frac{\sqrt{17}}{4}<x<-\frac{1}{4}+\frac{\sqrt{17}}{4}$$

To make the graph of the solution, just draw the number line from Figure 13.2 and darken in the solution interval, so you end up with something like Figure 13.3.

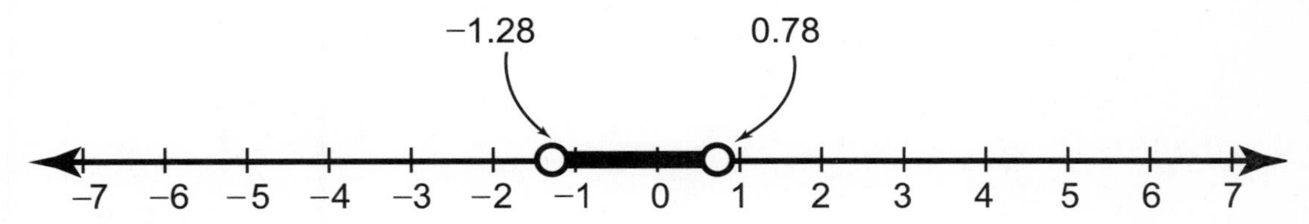

Figure 13.3

The solution graph for the inequality $2x^2 + x - 2 < 0$.

You've Got Problems

Problem 5: Solve the inequality and graph the solution $2x^2 + 5x - 3 \geq 0$.

The Least You Need to Know

- When you complete the square, the leading coefficient must be 1.
- Completing the square requires you to take half of the x coefficient and square it; this number should be added to both sides of the equation.
- The quadratic formula is $x = \dfrac{-b \pm \sqrt{b^2 - 4ac}}{2a}$.
- The discriminant correctly predicts how many solutions a quadratic equation will have.

Chapter 14

Solving High-Powered Equations

In This Chapter

- Solving cubic and higher-degree equations
- Factoring large polynomials
- Calculating roots
- Finding rational and irrational solutions

It wasn't long ago that you were cutting your mathematical teeth (they're the little ones, next to your bicuspids) on linear equations. Then, last chapter, I upped the ante, and showed you how to solve equations that were exactly 1 degree higher. Surprisingly, the techniques had very little in common, so you might enter this chapter with a little trepidation. Are you wondering, "If second-degree equations were so drastically different than first-degree equations, how different are third-degree equations going to be? Am I going to have to learn a new language (something as weird and antiquated as Latin or perhaps as complicated as dolphin) to figure these things out?"

I have good news for you; the answer is no. Although I will introduce a few new concepts in this chapter, you're not actually going to have to learn any new step-by-step methods. In fact, the techniques you use to solve third-degree equations are the same ones you'll use to solve fourth-, fifth-, sixth-, and fiftieth-degree equations, not that you'll see many of those. Even better, you'll use tools you've already mastered, like synthetic division, to bring high-powered equations to their knees.

There's No Escaping Your Roots!

To understand how to solve equations with larger exponents, I need to remind you how you solved quadratic equations. Specifically, I want to look at the factoring technique you used in Chapter 13 more closely.

Let's say you had some quadratic equation set equal to 0, and you were able to factor it like this:

$$(x - a)(x - b) = 0, \text{ where } a \text{ and } b \text{ are real numbers}$$

Remember what you did then? You set each individual factor equal to 0 to get the solutions (which you can also call the *roots* of the equation):

$$x - a = 0 \text{ or } x - b = 0$$

$$x = a \text{ or } x = b$$

There's an important relationship between the factors of a polynomial and its roots that this example illustrates: If $(x - a)$ is a factor of a polynomial, then $x = a$ is a root, or solution, when that polynomial is set equal to 0.

Kelley's Cautions

The solutions of an equation are also called the roots.

Critical Point

Precalculus courses focus even more on calculating polynomial roots and graphing polynomials of higher degree. For now, in algebra, we'll satisfy ourselves with just solving the higher-degree equations.

So, minus all the techno mumbo jumbo, what does that mean for you? Basically, to find the solutions when any polynomials are set equal to 0, you'll just factor them as much as possible and set each factor equal to 0. Occasionally, when you can't factor any further, you may have to resort to the quadratic formula or completing the square, whichever you feel more comfortable with, since both will give you the right answer.

That's the game plan. However, there's one problem: You only know how to factor quadratics! If higher powers are involved (here I mean exponents, not divine intervention), how are you supposed to factor? Good question. That's the next thing I need to discuss with you.

Finding Factors

I have discussed factoring and factors for so long at this point that, believe it or not, most students actually forget what factors actually are. (You know what I'm saying; sometimes, if you repeat a word over and over and over enough times, it stops sounding like a real word at all.) So, allow me to conduct a 16-second review with you on factors.

A factor is something that divides evenly into something else. In other words, if a is a factor of b, then when you calculate the quotient $b \div a$, there is no remainder. Factors work the same way with polynomials. Let's say $(x - m)$ is a factor of some big polynomial $ax^4 + bx^3 + cx^2 + dx + e$; if that's true, then the quotient $(ax^4 + bx^3 + cx^2 + dx + e) \div (x - m)$ will have no remainder.

Before we start actually factoring mammoth polynomials, I want to make sure you can determine whether or not a linear expression, like $x - m$, is a factor of a polynomial by applying synthetic division and ensuring that the remainder is 0.

Critical Point

If synthetic division is a little hazy in your mind, review it in Chapter 10. Remember, the rightmost number beneath the horizontal line represents the remainder.

Example 1: Demonstrate that $x - 3$ is a factor of the polynomial $x^3 - 6x^2 + 5x + 12$ and use that information to completely factor the cubic.

Critical Point

By testing to see if $x - 3$ is a factor in Example 1, you are also simultaneously testing to see whether or not 3 is a root of the equation $x^3 - 6x^2 + 5x + 12 = 0$.

Solution: If $x - 3$ is truly a factor, then dividing it into $x^3 - 6x^2 + 5x + 12$ using synthetic division should result in a remainder of 0. (Notice that $x - 3$ is linear, as all the factors will be in this chapter, so you can use synthetic, rather than long, division.) Don't forget that you should write 3, not –3, in the synthetic division box.

$$\begin{array}{r|rrrr} 3 & 1 & -6 & 5 & 12 \\ & & 3 & -9 & -12 \\ \hline & 1 & -3 & -4 & 0 \end{array}$$

The remainder's definitely 0, so $x - 3$ is a factor, and you can rewrite the polynomial in factored form: $(x - 3)(x^2 - 3x - 4)$. It's not *completely* factored yet, because the quadratic can, itself, be factored into $(x - 4)(x + 1)$ using the techniques you learned in Chapter 11.

Therefore, when fully factored, the polynomial $x^3 - 6x^2 + 5x + 12$ can be rewritten as $(x - 3)(x - 4)(x + 1)$; in fact, those three factors can be written in any order and the answer is still correct, thanks to the commutative property of multiplication.

You've Got Problems

Problem 1: Demonstrate that $x + 2$ is a factor of the polynomial $2x^3 + 5x^2 - 8x - 20$ and use that information to completely factor the trinomial.

Baby Steps to Solving Cubic Equations

I really get a kick out of the movie *What About Bob?* starring Bill Murray and Richard Dreyfuss. Murray is a needy patient in the therapy of mental health guru Dreyfuss, and greatly benefits from the doctor's philosophy of "baby steps." Paraphrased, this philosophy entails looking at a complex problem as a series of small, easily accomplished (baby) steps, rather than one giant leap from start to finish.

In keeping with this fictional, but useful, therapeutic technique, I have inserted a baby step into the process of solving high-powered equations. Whereas in Example 1 and Problem 1, you were required to factor a polynomial, now I want you to both factor a polynomial and solve an equation containing that polynomial. There's not a lot of difference between the two goals, but if not for this baby step, you might trip over the next thing I need to teach you.

As you solve these problems, I want you to cement in your mind the relationship between factors and roots within a high-powered equation. Remember, if a is a root of the equation, then $(x - a)$ is one of the polynomial's factors.

Example 2: Given that 5 is a root of the equation $x^3 + x^2 - 22x - 40 = 0$, use that information to find the other two roots.

Critical Point

A polynomial equation has up to n rational roots, where n is the degree of the polynomial. It may have fewer, but it cannot have more.

Solution: If 5 is a root of the equation, then $(x - 5)$ must be a factor. As you apply synthetic division to factor it out, notice that the number in the box matches the root you're given. In Chapter 10, I told you to put a in the synthetic division box when you're factoring out $(x - a)$, and now you know why: You're actually writing the corresponding root up there.

$$\begin{array}{r|rrrr} 5 & 1 & 1 & -22 & -40 \\ & & 5 & 30 & 40 \\ \hline & 1 & 6 & 8 & 0 \end{array}$$

You can now rewrite the equation and finish factoring the polynomial.

$$(x - 5)(x^2 + 6x + 8) = 0$$

$$(x - 5)(x + 4)(x + 2)= 0$$

Set each factor equal to 0 to discover that the roots of this equation are –4, –2, and 5.

You've Got Problems

Problem 2: Given that –2 is a root of the equation $x^3 + 10x^2 + 7x - 18 = 0$, use that information to find the other two roots.

Calculating Rational Roots

Does it feel like we've been cheating so far in this chapter? To be completely honest, we have been cheating a little bit. The goal of the entire chapter is to solve, or find the roots of, high-powered equations, and in every example so far, I am giving you part of the answer. If you didn't notice, I gave you one of the roots or factors every time, and told you to find the others based on it.

The process I taught you requires that you apply synthetic division using a known root or factor. But what happens when there's no one around to tell you one of the answers, giving you that kick start you need, that seed number for synthetic division that breaks the bigger, uglier polynomial into something like a quadratic that you can factor unaided?

Luckily, you've got a tool named the *rational root test.* This handy little test generates a list of all the numbers that could *possibly* be the rational roots of the equation; that's important because the roots you'll use to break down the polynomial equations need to be rational in order to apply the synthetic division technique you've been using so far in this chapter.

Basically, the rational root test is like a criminal lineup at the local police precinct. If you're a witness to a crime, the police may round up a bunch of thugs that loosely match your description of the guilty party, and ask you to identify the actual criminal from among the persons in that lineup. Not all of the people there are guilty of the crime (at least not the crime you witnessed, anyway), so it's your job to identify the responsible party or parties in the much larger group of possible suspects.

Talk the Talk

The **rational root test** uses the leading coefficient and constant of a polynomial to generate a list of all possible rational roots for a polynomial equation. Unfortunately, the list is usually much longer than the actual list of rational roots, so some trial and error is necessary until you identify an actual root in the list.

Back to the (slightly more boring) world of math. To round up all the possible rational root suspects of an equation, the rational root test looks at the leading coefficient and constant of a polynomial. More specifically, it looks at the factors of those numbers. It then lists all of the possible fractions of the form $\pm\frac{C}{L}$, where C is a factor of the constant and L is a factor of the leading coefficient. Sound strange? Wait until you see how it works in Example 3; it's not so hard.

Once all of the possible suspects are identified, it's time to separate the guilty (roots) from the innocent (nonroots). The best way to do this is synthetic division; just give each root a turn inside that synthetic division box and see which results in a remainder of 0. Once you spot that telltale remainder, tell the local police chief to slap on the cuffs, and then go back to the lineup, until all of the guilty suspects are apprehended.

Example 3: Identify all the rational roots for each equation.

(a) $2x^3 - 9x^2 - 8x + 15 = 0$

Solution: List all the factors of the leading coefficient, 2, and the constant, 15:

Factors of 2: 1, 2

Factors of 15: 1, 3, 5, 15

The rational root test tells you that all the positive and negative fractions you can create whose numerator is either 1, 3, 5, or 15 and whose denominator is either 1 or 2 are possible rational roots of the equation. The easiest way to make the list is to take each possible numerator and write it over each possible denominator, like so:

$$\pm\frac{1}{1}, \pm\frac{1}{2}, \pm\frac{3}{1}, \pm\frac{3}{2}, \pm\frac{5}{1}, \pm\frac{5}{2}, \pm\frac{15}{1}, \pm\frac{15}{2}$$

Yikes! There are 16 possible suspects (8 positive and 8 negative), and only a few of them are actually guilty of the crime of Criminal Activity in the Third Degree (Equation). Try them one at a time in synthetic division until you get a remainder of 0. Why not start with –1, for no good reason at all, except that it's at the front of the list.

$$\begin{array}{r|rrrr} -1 & 2 & -9 & -8 & 15 \\ & & -2 & 11 & -3 \\ \hline & 2 & -11 & 3 & 12 \end{array}$$

That didn't work; the remainder was 12. Oh well, one suspect cleared and 15 to go. This time, test to see if its opposite, 1, is a dirty, rotten root.

Critical Point

You won't always hit an actual root on your first or second stab at synthetic division. Just stick with it until you find something that works. However, if your instructor allows you to use graphing calculators, here's a trick: Test roots that appear to be x-intercepts of the polynomial's graph.

1⌋	2	–9	–8	15
		2	–7	–15
	2	–7	–15	0

Success! That means $x - 1$ is a factor, and the equation can be rewritten in factored form.

$$(x - 1)(2x^2 - 7x - 15) = 0$$

$$(x - 1)(2x + 3)(x - 5) = 0$$

The roots of the equation are $-\frac{3}{2}$, 1, and 5.

(b) $x^4 - 3x^3 - 19x^2 + 27x + 90$

Solution: The leading coefficient only has a factor of 1, but the constant's factors are 1, 2, 3, 5, 6, 9, 10, 15, 18, 30, 45, and 90. So, the possible rational roots are:

$$\pm1, \pm2, \pm3, \pm5, \pm6, \pm9, \pm10, \pm15, \pm18, \pm30, \pm45, \text{ and } \pm90$$

(Each possible root has a denominator of 1.) Once again, start at the beginning of the list. If you synthetically divide –1, 1, –2, and 2, you get remainders, but –3 will be a root.

–3⌋	1	–3	–19	27	90
		–3	18	3	–90
	1	–6	–1	30	0

So, the polynomial can be factored as $(x + 3)(x^3 - 6x^2 - x + 30)$, but an ugly trinomial remains, so you're not done synthetically dividing yet—back to the list of suspects. You don't have to try any of the roots that already failed (they still won't work), but when you test new possible roots, use the new, cubic polynomial, rather than the original quartic. It turns out that 3 is also a root.

Kelley's Cautions

If you don't use the 1-degree smaller *result* of synthetic division each time you find a root (rather than the *original* polynomial) when solving Example 3(b) and other polynomials of degree 4 and higher, you'll never get it down to a quadratic you can factor.

$$\begin{array}{r|rrrr} 3 & 1 & -6 & -1 & 30 \\ & & 3 & -9 & -30 \\ \hline & 1 & -3 & -10 & 0 \end{array}$$

Here's the latest factored form of the quartic polynomial: $(x + 3)(x - 3)(x^2 - 3x - 10)$. Finally, the nonlinear factor is a quadratic, and you can break it up without synthetic division: $(x + 3)(x - 3)(x - 5)(x + 2)$. So, the roots of the equation are –3, –2, 3, and 5.

You've Got Problems

Problem 3: Identify all four rational roots of the polynomial $4x^4 + 4x^3 - 9x^2 - x + 2 = 0$.

What About Imaginary and Irrational Roots?

So far, these high-powered equations have worked out nicely. In the end, we're always getting a total number of rational roots equal to the degree of the polynomial, so you might be tempted to assume that will always be true. Unfortunately, it's not. Occasionally, you'll end up with some irrational numbers (that contain radicals which simply refuse to be simplified) or imaginary numbers.

How'd You Do That?

According to the *Fundamental Theorem of Algebra,* a polynomial of degree *n,* when set equal to 0, will have *exactly n roots.* They may be real, imaginary, rational, or irrational, but their number will always match the degree of the equation. Note that imaginary roots will always come in pairs (that are conjugates of one another) and double roots count as two roots.

You'll just be tootling along, trying to locate rational roots, breaking polynomials up into smaller factors with synthetic division, just as you have all chapter long, and all of a sudden, you'll end up with a quadratic that just doesn't factor, meaning you've encountered either an imaginary or irrational root. No problem—just apply either the quadratic formula or completing the square and simplify your answer.

Example 4: Find all the roots of each equation.

(a) $x^3 - 8x^2 + 4x - 32 = 0$

Solution: According to the rational root test, this is the list of possible rational root suspects:

$$\pm 1, \pm 2, \pm 4, \pm 8, \pm 16, \pm 32$$

Believe it or not, nothing qualifies as a root until you try 8.

$$\begin{array}{r|rrrr} 8 & 1 & -8 & 4 & -32 \\ & & 8 & 0 & 32 \\ \hline & 1 & 0 & 4 & 0 \end{array}$$

Factored, the equation now looks like $(x - 8)(x^2 + 4) = 0$. Notice that $x^2 + 4$ is the *sum* of perfect squares (not the *difference* of perfect squares) and cannot be factored further. However, you can still set that factor equal to 0 and solve the resulting equation:

$$\begin{aligned} x^2 + 4 &= 0 \\ x^2 &= -4 \\ \sqrt{x^2} &= \sqrt{-4} \\ x &= \pm 2i \end{aligned}$$

So, the roots of the equation are 8, $2i$, and $-2i$. Notice that there are three of them (which matches the degree of the polynomial) and that the imaginary roots are a conjugate pair.

(b) $x^4 + 2x^3 - 10x^2 - 5x + 12 = 0$

Solution: Of the possible rational roots, ±1, ±2, ±3, ±4, ±6, ±12, only –4 and 1 are valid. Once you apply synthetic division for both roots, you end up with $(x - 1)(x + 4)(x^2 - x - 3) = 0$. The remaining quadratic is prime (unfactorable), so when you set it equal to 0, you'll have to either complete the square or apply the quadratic formula.

$$x = \frac{1 \pm \sqrt{13}}{2}$$

So, the roots of the equation are –4, 1, $\frac{1}{2} + \frac{\sqrt{13}}{2}$ (approximately 2.303), and $\frac{1}{2} - \frac{\sqrt{13}}{2}$ (approximately –1.303). Once again, there is a total of four roots, two rational ones and two that are irrational.

You've Got Problems

Problem 4: Find all the roots of the equation $2x^3 - 5x^2 + 4x - 21 = 0$.

The Least You Need to Know

- The solutions of an equation are also called its roots.
- If $(x - b)$ is a factor of a polynomial, then b will be a root of the equation created when the polynomial is set equal to 0.
- The rational root test lists all possible rational roots for a polynomial equation based on its constant and leading coefficient.
- A polynomial equation of degree n always has exactly n roots, although some may be imaginary or irrational.

Part 5
The Function Junction

In his book *Prey*, Michael Crichton describes the horror of what can happen when nanomachines (tiny little robots the size of molecules) go out of control and wreak havoc in a murderous rampage. In this part, you'll learn that mathematics is powered by similar little nanomachines called functions. The best part about them is that they're just theoretical, and pose no physical danger to you whatsoever, unless of course you drop your algebra book on your foot or something. Then they hurt like crazy.

Chapter 15

Introducing the Function

In This Chapter

- Differentiating between functions and relations
- Combining functions
- Designing inverse functions
- Understanding piecewise-defined functions

Brace yourself. It's time to change algebraic gears for a moment, and you might feel a slight lurching in your stomach. Until now, I have spoken only of expressions and equations, and although you will continue to see both, you're going to see a lot more of something called the *function* (especially in later math courses).

This concept is so important, in fact, that I have dedicated this and the next chapter to the study of functions. They're not a whole lot different than regular old equations or expressions, but since they are so essential, I want to give you some time to get used to them.

Getting to Know Your Relations

Before I tell you what a function is, I need to step back for a moment and discuss vending machines. As a one-time college student, I fully appreciate the importance of these handy, rectangular, poor man's concierges. The rickety, but trustworthy, dorm vending machine (helpfully labeled "Tasty Snacks" and centrally located in the TV lounge) didn't mind if it was 2 A.M. when you came looking for peanut butter cheese crackers. For a mere 50 cents and a press of the right button, you were seconds away from cellophane-wrapped peanut buttery goodness (and inexplicably orange fingers).

In the mathematical world, there are an infinite number of little vending machines called *relations*, rules that swallow a number input (like a vending machine swallows a coin and waits for a button press) and output another number (like a snack machine dispenses both treats for your taste buds and cholesterol directly to the lining of your artery walls).

Basically, the job of a relation is to define a *relationship* between the inputs and outputs for the vending machine. In other words, a relationship provides the answer to "What will you give me if I give you this?" The simplest way, therefore, to define a relation is to list the inputs and the corresponding outputs as ordered pairs, like so:

$$s: \{(1,3), (2,7), (3,-1), (4,9), (5,-1), (5,0)\}$$

This relation, called s, tells you that an input of 1 results in an output of 3, thanks to the included pair (1,3). If you walk up to the s vending machine and pop a 1 in the coin slot, out comes a 3 automatically (possibly coated in peanut butter and/or containing nougat). Mathematically you write $s(1) = 3$, which is read "s of 1 equals 3." Similarly, you could write $s(2) = 7$, $s(3) = -1$, and so forth.

However, something buggy happens to the s vending machine when you input 5—did you notice? According to the definition of s, $s(5) = -1$ and $s(5) = 0$. In other words, if you plug 5 into the relation, you might get one answer, and you might get the other answer. You might even get both (if you shook the vending machine hard enough, I suppose).

Talk the Talk

A **relation** is a machine-like rule that accepts input numbers and provides corresponding output numbers. If every input has only one corresponding output, then the relation is classified as a **function**.

In the mathematical world, a small fraction of the relations receive a special commendation for good and reliable service; those relations are called *functions*. However, s will never receive this commendation, because the sole requirement for a relation to be classified as a function is that every one of its inputs must always result in a single, predictable output.

Since s fails that litmus test for the input 5, it will always be a second-class citizen, a nonfunction.

How'd You Do That?

Consider the relation r as defined below.

$$r: \{(-2,4), (-1,9), (0,4)\}$$

Is r a function? There is something troubling about r; notice that both $r(-2)$ and $r(0)$ are equal to 4. In other words, if you input either −2 or 0, you'll get the same thing out: 4. That is allowed for functions; remember, in order to classify as a function, each input must result in only one corresponding output, and that's true for r. If you input −2 or 0, you'll always get out 4. An input of −1 always results in an output of 9. So this relation passes the function qualification. You're looking for a single, predictable output for each input.

There's no rule saying that functions can't provide the same output for multiple inputs. However, there is a special name for functions whose inputs each have a *unique* output—they are called *one-to-one functions*.

Functions and relations are rarely written as lists of ordered pairs, because most accept an infinite number of inputs, and writing out an infinite list of numbers is generally considered not the best use of one's time (and can sometimes lead to carpal tunnel syndrome, a decrease in your attractiveness to the opposite sex, and/or hives).

Usually, relations are written in terms of the input, usually x, like this:

$$h(x) = 3x + 4$$

In fact, h is actually a function, because no matter what you substitute in for x, you'll always get a single, predictable, corresponding answer. For instance, given the input –3, you always get an output of –5.

$$\begin{aligned} h(-3) &= 3(-3) + 4 \\ &= -9 + 4 \\ &= -5 \end{aligned}$$

No matter how many times you plug in $x = -3$, you'll get out –5 in the end; the same is true for any input.

I don't want to get mired down in testing relations to see if they're functions right now, so even if you're not sure how I was so positive that h was a function, don't worry. In Chapter 16, I'll introduce you to the vertical line test, a quick method to easily determine whether or not a relation is a function. It's sort of like a pregnancy test, only without the messiness of, well, you know. For the rest of this chapter, I want to focus on the basic sorts of things you'll be asked to do with functions.

Operating on Functions

Every time I've introduced something new (fractions, matrices, polynomials, and radicals, to name just a few), it seems like the first thing I show you how to do is add, subtract, multiply, and divide those new things; functions will be no different. However, since functions, are made up of things like fractions, polynomials, and radicals, performing operations on them will seem oddly familiar, like that person you saw at the grocery store the other day who you just *swore* you recognized, but couldn't be sure who they were.

The only truly new thing you'll learn as you combine functions with these operations is the notation to use as you do so.

Example 1: Given the functions $f(x) = x^2 - 4x + 3$ and $g(x) = x - 1$, find (a) $(f + g)(x)$; (b) $(f - g)(x)$; (c) $(fg)(x)$; and (d) $\left(\frac{f}{g}\right)(x)$. In addition, evaluate each function at $x = 2$.

Solution: Basically, you're asked to come up with four new functions, based on two given functions (f and g) and the four basic operations. Also, once you've found these new functions, you're supposed to plug $x = 2$ into each to figure out the corresponding output.

(a) Don't even worry about the fact that they're functions; just add the quadratic function f to the linear function g by combining like terms.

$$\begin{aligned}(f + g)(x) &= (x^2 - 4x + 3) + (x - 1)\\ &= x^2 - 3x + 2\end{aligned}$$

Don't forget to evaluate $(f + g)(2)$, as requested by the problem.

$$(f + g)(2) = (2)^2 - 3(2) + 2 = 0$$

(b) You should distribute the –1 through the quantity $(x - 1)$ before combining like terms.

$$\begin{aligned}(f - g)(x) &= (x^2 - 4x + 3) - (x - 1)\\ &= x^2 - 4x + 3 - x + 1\\ &= x^2 - 5x + 4\\ (f - g)(2) &= (2)^2 - 5(2) + 4 = -2\end{aligned}$$

Critical Point

I evaluated $f(2)$ at the end of each part of Example 1, but that's not the only way to do these problems; take part (a) as an example. Instead of plugging 2 into the new function $(f + g)(x)$, you can plug 2 into each of the original functions, $f(x)$ and $g(x)$, and then just add the results together.

$$f(2) = 2^2 - 4(2) + 3 = -1 \qquad g(2) = 2 - 1 = 1$$

Notice that $-1 + 1 = 0$, which matches the output $(f + g)(2)$. The same process works for the other operations as well.

(c) Multiply the polynomials by distributing each term of f through each term of g, one at a time.

$$\begin{aligned}(fg)(x) &= (x^2 - 4x + 3)(x - 1)\\ &= x^3 - x^2 - 4x^2 + 4x + 3x - 3\\ &= x^3 - 5x^2 + 7x - 3\end{aligned}$$

$$(fg)(2) = 2^3 - 5(2)^2 + (7)2 - 3 = -1$$

(d) Since $(x - 1)$ is a factor of $x^2 - 4x + 3$, you can simplify the fraction in $\left(\frac{f}{g}\right)(x)$ using long or synthetic division.

$$\begin{aligned}\left(\frac{f}{g}\right)(x) &= \frac{x^2 - 4x + 3}{x - 1}\\ &= x - 3\end{aligned}$$

$$\left(\frac{f}{g}\right)(2) = 2 - 3 = -1$$

Critical Point

I'll explore rational functions like $\left(\frac{f}{g}\right)(x)$ from Example 1(d) in greater detail in Chapters 17 and 18. There are much better ways to simplify fractions like that without resorting to long or synthetic division.

You've Got Problems

Problem 1: Given the functions $h(x) = x^2 + 7x - 5$ and $k(x) = 4x - 3$, evaluate $(h + k)(-1)$, $(h - k)(-1)$, $(hk)(-1)$, and $\left(\frac{h}{k}\right)(-1)$; afterward, put them in order from least to greatest.

Composition of Functions

Sometime during college, my friends Matt and Chris and I decided to hold a contest. It was the sort of intellectual competition that's only dreamed up by those residing in the hallowed halls of academia, a clash of wit and cunning designed while our minds were at peak functioning power. Basically, we wanted to see who could stuff the most Cheerios in his mouth before he gagged or asphyxiated, drawing a legion of tan breakfast O's into his lungs.

Chris won. I don't remember the absurd number of Cheerios he was able to jam into his maw (and I don't recommend trying this at home), but I admired his dogged determination not to lose the contest, although nothing was at stake (except bragging rights, of course). Even today, I believe some Cheerios reside in his sinus cavities as a de facto trophy of the event, accidentally sucked in when he started gagging.

If anything, the Cheerio Jam taught me two major lessons. First, college students don't always understand what's important in the "big picture," and second, sometimes it's fun to jam more into something than you usually do.

So far this chapter, you've only input real numbers into functions, normal, bite-sized mouthfuls that the functions chewed up easily and spit out as outputs. Now, however, the inputs are going to get bigger; rather than single numbers, we're going to feed the functions *entire other functions* to see what happens, in a process called *composition of functions.*

Talk the Talk

The process of inputting one function into another is called the **composition of functions,** and can be denoted with a circle

Kelley's Cautions

If a function $f(x)$ contains multiple x's, and you're trying to create the composition $f(g(x))$, make sure to plug $g(x)$ into *every one of those* x variables in f.

This is not the first time you've substituted an entire expression for a variable, rather than just a number; in fact, you did it as recently as Chapter 8, during the substitution method. However, composition of functions differs a bit as it has its own notation.

The intent to plug a function $g(x)$ into another function $f(x)$ is indicated by the notation $(f \circ g)(x)$, read "f composed with g of x" or "f circle g of x." That little round thing in there is, indeed, a circle—a new operator that joins the ranks of +, –, ·, and ÷ that basically means "plug the right function in for the left function's variable." Sometimes, function composition is written explicitly, like this: $f(g(x))$ (read "f of g of x"), sort of like $f(2)$, except instead of plugging 2 into the function, you plug in $g(x)$ instead.

Example 2: If $g(x) = 2x - 1$, and $h(x) = x^2 + 3x$, find $(h \circ g)(x)$.

Solution: This means the same thing as $h(g(x))$; basically, you should plug function g, which is $2x - 1$, into every x-spot in $h(x)$.

$$h(x) = x^2 + 3x$$

$$h(2x - 1) = (2x - 1)^2 + 3(2x - 1)$$

Don't get confused because both functions contain x's; just think of the x's as placeholders where eventual values will go. In this case, all of the x's in h receive the value $2x - 1$. Finish by simplifying the function.

$$\begin{aligned} h(2x - 1) &= 4x^2 - 4x + 1 + 6x - 3 \\ &= 4x^2 + 2x - 2 \end{aligned}$$

Once you get over the initial unease about replacing a variable with an expression containing a variable, you're ready to start composing larger and larger functions, until you accidentally gag, drawing numbers and variables into your sinus cavity, like Chris.

You've Got Problems

Problem 2: If $f(x) = x^2 + 5$ and $g(x) = \sqrt{x - 2}$, determine the values of the following:

(a) $(f \circ g)(x)$

(b) $(g \circ f)(x)$

Inverse Functions

So far in this book, you've encountered a few functions with a unique superpower: unraveling other functions. Consider the equation:

$$(x - 2)^2 = 5$$

Even though you could use the quadratic formula to solve this, the easiest way to reach a solution would be to take the square root of both sides to get:

$$\begin{aligned} \sqrt{(x-2)^2} &= \pm\sqrt{5} \\ x - 2 &= \pm\sqrt{5} \\ x &= 2 \pm \sqrt{5} \end{aligned}$$

In Chapter 12, you learned that a root with index n cancels out an exponent of n. That is because they are *inverse functions.*

Inverse functions balance one another out in the mathematical world; they represent the eternal struggle between yin and yang, push and pull, right and left, really fun (like Disney's Magic Kingdom and MGM Studios in Florida) and extremely boring (like Disney's Epcot Center—"Hey, who wants to watch another 45-minute movie on the troposphere?").

Defining Inverse Functions

Mathematically speaking, if you compose inverse functions with one another, they cancel each other out. In other words, if $f(x)$ and $g(x)$ are inverses of one another, then:

$$f(g(x)) = g(f(x)) = x$$

See how they canceled each other out? If you plug f into g or g into f, you get the same thing, x (the input of the innermost function). To denote that $g(x)$ is the inverse of $f(x)$, you would write $g(x) = f^{-1}(x)$ (read "g of x is equal to f inverse of x," or "g of x is the inverse of f of x"). You could also correctly write $f(x) = g^{-1}(x)$. Even though inverse function notation *looks* like a negative exponent, it's not, so don't start thinking about taking the reciprocal of f or anything.

Talk the Talk

Inverse functions cancel each other out when composed with one another. In other words, $(f \circ g)(x) = (g \circ f)(x) = x$. The inverse of the function $f(x)$ is written $f^{-1}(x)$.

Critical Point

Only one-to-one functions (functions whose inputs each have unique outputs) possess inverses. For now, when I ask you to find the inverse of a function, assume that you're allowed because the function is one-to-one; you'll learn a technique to actually determine for yourself if a function is one-to-one in Chapter 16.

Inverse functions also have another cool property: If you reverse the ordered pair in one function, you get an ordered pair in the inverse function. Therefore, if $f(a) = b$, then $f^{-1}(b) = a$. If that doesn't make sense to you, think of it this way. If you plug the number 2 into some function f and get out -7, then you automatically know that plugging -7 into f's inverse function will give you 2.

Example 3: Demonstrate that $f(x) = 4x + 3$ and $g(x) = \frac{1}{4}x - \frac{3}{4}$ are inverses of one another.

Solution: When you compose f with g and g with f, you should end up with just x in both cases.

$$\boxed{f(g(x))}$$

$$= f\left(\frac{1}{4}x - \frac{3}{4}\right)$$

$$= 4\left(\frac{1}{4}x - \frac{3}{4}\right) + 3$$

$$= x - 3 + 3$$

$$= x$$

$$\boxed{g(f(x))}$$

$$= g(4x + 3)$$

$$= \frac{1}{4}(4x + 3) - \frac{3}{4}$$

$$= x + \frac{3}{4} - \frac{3}{4}$$

$$= x$$

You've Got Problems

Problem 3: Demonstrate that $f(x) = 3x - 5$ and $g(x) = \frac{x+5}{3}$ are inverse functions.

Calculating Inverse Functions

The process you use to actually come up with an inverse function $f^{-1}(x)$ for a given function $f(x)$ is very simple. Just follow these steps:

1. **Rewrite $f(x)$ as y.** This gives you two variables in the equation, instead of just x.
2. **Switch x and y.** Wherever you see an x, change it to y and vice versa. This is how you force the function into an inverse—remember, a function and its inverse will contain ordered pairs that are the reverse of one another; you're actually forcing them to reverse in this step.
3. **Solve for y.** Isolate y on one side of the equation.
4. **Rewrite y as $f^{-1}(x)$.** You're all finished!

That's all I'm going to say about inverse functions. Now that you can find them for yourself and verify that they are inverses, I am satisfied. There are more details about inverse functions (including technical details such as restrictions) that you'll learn in later courses, but I omit them here because they are unnecessarily complicated to discuss right now.

Example 4: If $f(x) = \sqrt{x+2}$, find $f^{-1}(x)$.

Solution: Rewrite $f(x)$ as y.

$$y = \sqrt{x+2}$$

Switch the *x*'s and *y*'s.

$$x = \sqrt{y+2}$$

Solve for *y*. To eliminate the radical sign, square both sides of the equation. (You don't add a "±" sign; you only do that when you take the square *root* of both sides of an equation.)

$$x^2 = y + 2$$

$$x^2 - 2 = y$$

Rewrite *y* as $f^{-1}(x)$. You may also swap the sides of the equal sign to get $f^{-1}(x)$ on the left, if you wish.

$$f^{-1}(x) = x^2 - 2$$

You've Got Problems

Problem 4: If $g(x) = 7x - 3$ find $g^{-1}(x)$.

Bet You Can't Solve Just One: Piecewise-Defined Functions

I dated a few people before I met my wife. When I was very young, I figured that as soon as I found somebody that I liked (and who liked me) that I'd probably just go ahead and get married. Little did I know that dating was actually an exercise in determining what you can and can't put up with in a partner.

Consider your own dating history. One person's very nice, but too wishy-washy. Another person's very self-assured but too possessive. Yet another person has a repulsive personality, but doesn't mind the way your feet smell. Basically, the entire time you're dating, you're really designing a wish list for the perfect mate: He or she has to be self-assured, but not pushy; nice but not saccharinely sweet; and also able to put up with the smell of your toes when you kick off wet shoes. Luckily, someone eventually comes around who meets just about all of those qualities (face it, your feet aren't going to win you any prizes), and you take the plunge.

Things are a little different with functions; you can actually design the perfect function—one that meets any qualifications you set. In fact, if you like certain parts of different functions, you can take the parts you like, sew them together, and design a new function that's a patchwork quilt of your favorite function pieces called a *piecewise-defined function*.

The smaller, component functions are surrounded by one big brace on the left and look something like this:

$$f(x)=\begin{cases} g(x), x \le a \\ h(x), x > a \end{cases}$$

The function f is actually made up of two other functions, g and h. You actually use g whenever your input is less than or equal to a, but turn around and plug into h whenever your input is greater than a. You can even have more than two little functions sewn together in a piecewise-defined function, as illustrated by Example 5.

Talk the Talk

A **piecewise-defined function** is made up of two or more functions that are restricted according to input; each individual function that makes up the piecewise-defined function is valid for only certain x input values.

Kelley's Cautions

Piecewise-defined functions must still meet the general function requirement: Each input must result in only one corresponding output. To avoid violating this requirement, make sure the input restrictions (the little inequalities written next to the smaller function components) don't overlap.

For instance, consider $h(x)$ defined here:

$$h(x)=\begin{cases} x^2-3, & x<4 \\ 2x+1, & x \ge 0 \end{cases}$$

Notice that an input of $x = 1$ would qualify to be plugged into both functions, since $1 < 4$ and $1 \ge 0$. Therefore, $h(1) = 1^2 - 3 = -2$ and $h(1) = 2(1) + 1 = 3$. Because this (and other) inputs give you more than one corresponding output, h cannot be a function.

Example 5: Given $m(x)$ as defined here, evaluate: (a) $m(-3)$, (b) $m(9)$, and (c) $m(2)$.

$$m(x)=\begin{cases} 4x-x^2, & x \le -1 \\ \sqrt{x+7}, & -1 < x < 5 \\ x-16, & x \ge 5 \end{cases}$$

Solution: Look at the input restrictions to determine which function to plug each number into.

(a) Since $-3 \le 1$, you should plug $x = -3$ into the top function.

$$\begin{aligned} m(-3) &= 4(-3) - (-3)^2 \\ &= -12 - 9 \\ &= -21 \end{aligned}$$

(b) Plug $x = 9$ into the function $x - 16$, since $9 \ge 5$.

$$\begin{aligned} m(9) &= 9 - 16 \\ &= -7 \end{aligned}$$

(c) Because 2 falls within the interval bounded by –1 and 5 ($-1 \le x \le 5$), you should plug it into the middle function.

$$\begin{aligned} m(2) &= \sqrt{2+7} \\ &= \sqrt{9} \\ &= 3 \end{aligned}$$

You've Got Problems

Problem 5: Given the function $f(x)$ as defined below, put the values $f(-1)$, $f(1)$, and $f(3)$ in order from least to greatest.

$$f(x) = \begin{cases} 3x+4, & x \le 0 \\ x-x^2, & x > 0 \end{cases}$$

The Least You Need to Know

- A relation pairs inputs together with outputs.
- A function is a relation whose inputs each result in a single, corresponding output.
- When you plug one function into another, you are performing composition of functions.
- A function and its inverse, when composed together, cancel one another out.
- A piecewise-defined function is made up of multiple other functions saddled with input restrictions.

Chapter 16

Graphing Functions

In This Chapter

- Sketching graphs using brute force and finesse
- Characterizing functions based on their graphs
- Stretching, squishing, and moving graphs
- Determining the domain and range of functions

Even though you have factored and solved nonlinear equations for quite some time now, I haven't showed you their graphs yet. There's a reason for this: It may freak you out just a little bit. You're probably used to nice, straight lines, and now, all of a sudden, curves get thrown into the mix. It's a world-shaking change.

It's sort of like sitting across from your mom at the breakfast table one morning, when she mentions in an off-handed manner that she's been a crime-fighting superhero named The Silver Heron for, well, going on 12 years now. Your mom! The same kindly lady who'd brought cupcakes to your fourth-grade class when it was your birthday and tied empty bread bags around your feet before you put your boots on to go play in the snow so your socks would stay extra dry.

Of course, as cool as it would be to have a mom that could single-handedly thwart evildoers' plans to rule the world, it would take some getting used to. A few things would stay the same in your mind, I'm sure, but lots of other things would change. Undoubtedly, it would take some time to adjust.

Metaphorically speaking, that is the same kind of jump you're about to make with graphing. The kindly old graphs that used to contain bland, boring lines are about to house spectacular-looking curves. While some minor things about graphing (like the coordinate system) will stay the same, you're going to have to learn some new techniques to master all these new graphical foes. After all, Mom needs a sidekick.

Second Verse, Same as the First

Think back to your initial attempts at graphing for a moment; do you remember how you threw together your first linear graphs? You created a table, plugging in a few values for x to get the corresponding y-values so that you could plot (x,y) points in the coordinate plane. Even though you'll be graphing functions now instead of equations, you can use basically the same method to get an accurate graph.

You'll need to plug a bunch of x values into the function to get the corresponding $f(x)$ outputs. Then, all you do is plot the $(x, f(x))$ points on the coordinate plane. For instance, if you plug $x = 1$ into a function $f(x)$, you'll get the output $f(1)$, and you can then plot the point $(1, f(1))$ in the plane. Once you plug in enough x-values, you'll start to get a good sense of what the graph looks like.

Kelley's Cautions

Sometimes, when you plug in values for x, you won't get a valid, real number output. That's okay; it just means that the number you plugged in is not a valid input for the function. (To use terminology you'll learn later in the chapter, that x-value is not in the *domain*.) Just keep plugging in different x's until you find some that work.

Critical Point

The u-shaped graph of a quadratic is called a parabola.

Example 1: Sketch the graph of $f(x) = (x + 2)^2 - 4$ by creating a table.

Solution: Because you're in the land of nonlinear graphs, you'll have to plot more than two or three points to get a good idea of what the graph should look like. This takes some experimenting. If you plug in x-values that give either really large or really negative outputs, they're kind of hard to graph. Try other inputs, until you get some smaller results.

In the case of $f(x) = (x + 2)^2 - 4$, inputs less than -5 or greater than 1 begin to give you unreasonably large results; however, if you evaluate the function for the integers within the interval $-5 \le x \le 1$, as I have done in the table that follows, you get a nice look at the graph of f.

x	$f(x)=(x+2)^2-4$
-5	$f(-5)=(-3)^2-4=5$
-4	$f(-4)=(-2)^2-4=0$
-3	$f(-3)=(-1)^2-4=-3$
-2	$f(-2)=(0)^2-4=-4$
-1	$f(-1)=(1)^2-4=-3$
0	$f(0)=(2)^2-4=0$
1	$f(1)=(3)^2-4=5$

To graph the function, just plot the points (–5,5), (–4,0), (–3,–3), and so on, as illustrated in Figure 16.1.

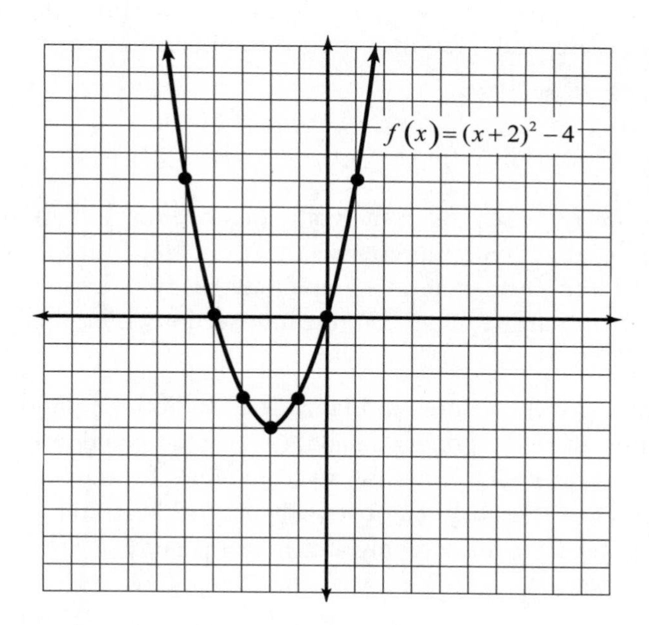

Figure 16.1

The graph of f(x) = (x + 2)² – 4, *with dots marking each point generated by the table of values.*

There's one thing I want to draw your attention to in Figure 16.1. This graph (unlike linear graphs) has two x-intercepts, at (–4,0) and (0,0). The x-values from those ordered pairs are called the *zeros* of the graph (since they represent the values of x for which $f(x)$ equals 0). In other words, the zeros of a function $f(x)$ are the *exact same thing* as the roots of the equation $f(x) = 0$.

Talk the Talk

The **zeros** of a function $f(x)$ are the x-values that make the equation $f(x) = 0$ true. Graphically speaking, the zeros are the x-intercepts of the function.

You've Got Problems

Problem 1: Use a table of values to graph $g(x) = -|x-5| + 3$.

Two Important Line Tests

Remember, predatory cats sleep a large portion of the day, so that's not a sign of weakness. Instead, you should ensure the large animal maintains its sharp eyes and keen reflexes; a loss in appetite or a general malaise toward food is often a symptom of a larger problem.

Wait a minute, I read the title of this section wrong. These are important *lion* tests, not *line* tests. My mistake. There are two major line tests in algebra, one involving vertical lines, and one horizontal lines; they're used to discern information about a function based solely on its graph.

The Vertical Line Test

The *vertical line test* allows you to look at the graph of a relation, and to immediately conclude whether or not that graph represents a function. Here's how it works. If you were to draw a vertical line *anywhere* on the graph, how many times would that line intersect the graph? If your answer is *at most* one intersection point, then the graph is a function.

Take a look at the three graphs pictured in Figure 16.2. In both graphs (a) and (b), there are a whole bunch of vertical lines you could draw that would intersect the graph in two places, even though I only show two such lines for each. (I made them dotted lines, to remind you that they're not actually part of the graph or anything—they're just little strips used for the test only, like litmus paper that's thrown away once the experiment's over.)

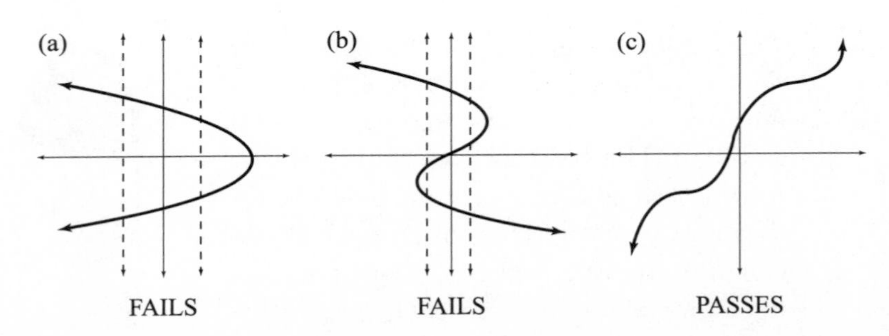

Figure 16.2

Graphs (a) and (b) fail the vertical line test. Only (c) gets passing marks, meaning that it's a function.

That's not true for graph (c), though. No matter where you draw vertical lines on that graph, they'll intersect at most at only one point, so that relation is definitely a function.

How'd You Do That?

The vertical line test is so simple and yet gives you very important information. Are you wondering why it actually works?

When you were first introduced to the coordinate plane in Chapter 5, you learned that all vertical lines look like the equation $x = c$, where c is a real number. When you draw vertical lines on a graph and find an intersection point, you're actually finding y-values to correspond with that x represented by that vertical line.

So, if a vertical line intersects a graph twice, then there are two y-values (or outputs) that can result for that x-value (or input), which is precisely what disqualifies a relation from being a function.

The Horizontal Line Test

Once you know that a relation's a function, you might want to take it one step further, to see if that function is one-to-one; this is accomplished with the *horizontal line test.* Remember, only one-to-one functions have inverses, so you shouldn't even bother trying to calculate the inverse of a function unless it first passes the horizontal line test.

The horizontal line test works a lot like the vertical line test did. Basically imagine a series of horizontal lines running across the graph. If you're positive that *any* horizontal line will intersect the graph just once *at most* (in either test, vertical and horizontal lines are allowed to miss the graph altogether; they just can't hit it multiple times), then the graph passes the test and is classified one-to-one.

To see the horizontal line test in action, let your eyes stroll to Figure 16.3, another trio of graphs currently undergoing the horizontal line test.

Talk the Talk

The **vertical line test** allows you to determine whether or not a relation is a function by examining its graph. The **horizontal line test** uses a similar technique to determine whether or not a function is one-to-one.

How'd You Do That?

Since horizontal lines represent y-values (outputs), they can't intersect the graph more than once if the graph is one-to-one. Otherwise, multiple x's would have the same output, violating the definition of one-to-one functions.

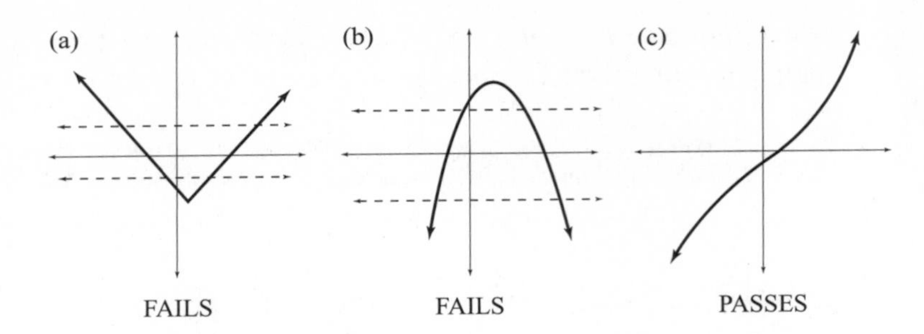

Figure 16.3

Only graph (c) passes the horizontal line test.

Once again, many lines (two of which I've drawn in there as dotted lines) will intersect at two places, so while graphs (a) and (b) remain functions (they still pass the vertical line test), they are not one-to-one functions.

You've Got Problems

Problem 2: In the last problem, you generated the graph of $g(x) = -|x-5|+3$. Can you verify that g is a function? If so, is it one-to-one? And do you think it would like this apple pie I made for dessert?

Determining Domain and Range

Up to this point, I have discussed the numbers that go into functions and the numbers that come out of them, respectively, as the inputs and outputs; however, it's time to use Official Math Terminology (drum roll, please). All of the numbers you're allowed to stick into a function as x collectively make up the *domain* of that function. Therefore, if a number belongs to the domain, it represents a valid input.

Kelley's Cautions

For the remainder of this book (and for the rest of algebra in general), we're just going to consider real number answers; complex numbers are really hard things to deal with, so from this point forward, let's pretend like they don't exist, okay? This is normal behavior for an algebra student; in fact, complex numbers are so tricky, you won't learn much more about them unless you become a math major in college!

Think about the function $f(x) = \frac{1}{x-2}$ It seems as though you can plug in any real number in the known universe for x, and that input will have a matching output.

However, there's one number that stubbornly won't work as an input: $x = 2$. When you try to calculate $f(2)$, you get this:

$$f(2)=\frac{1}{2-2}=\frac{1}{0}$$

That result does not compute! It violates one of the only rules governing the construction of fractions: You can never divide by 0. (The other two rules are: Never get fractions wet and never feed them after midnight.) Because $x = 2$ is not a valid input, it is the only real number not in the domain. (No other input causes a forbidden 0.)

Once you figure out the domain of a function, plugging in all of those inputs gives you a collection of values called the *range*, the set of valid function outputs. (Since the *domain* of a function is usually infinite, luckily, there are ways to figure out the range without having to plug in each input.)

Critical Point

The two major reasons that numbers are excluded from a function's domain are either (1) they cause 0 in the denominator of a fraction, or (2) they cause a negative number within a radical.

The best way to determine what, exactly, a function's domain and range are is to examine the graph and perform modified versions of the vertical and horizontal line tests. To figure out the domain, ask yourself this question: "If I draw vertical lines through the graph, which ones will intersect the graph, and which ones won't?" Any vertical lines that don't hit the graph represent x-values that aren't in the domain.

Talk the Talk

The **domain** of a function is the collection of numbers that can be plugged into it, the valid set of inputs. The **range** of a function is the collection of outputs.

Similarly, imagine lots of horizontal lines superimposed over the function. Any such lines that don't intersect the graph represent y-values that aren't in the range. Who knew that you could do so many useful things with horizontal and vertical lines?

Critical Point

In case you're wondering where in the world the graph of *f(x)* came from in Example 2, don't worry—you'll know how to graph that and many other functions by the end of the chapter.

Example 2: Determine the domain and range of the function $f(x)=\sqrt{x+3}-1$, whose graph is pictured on the next page.

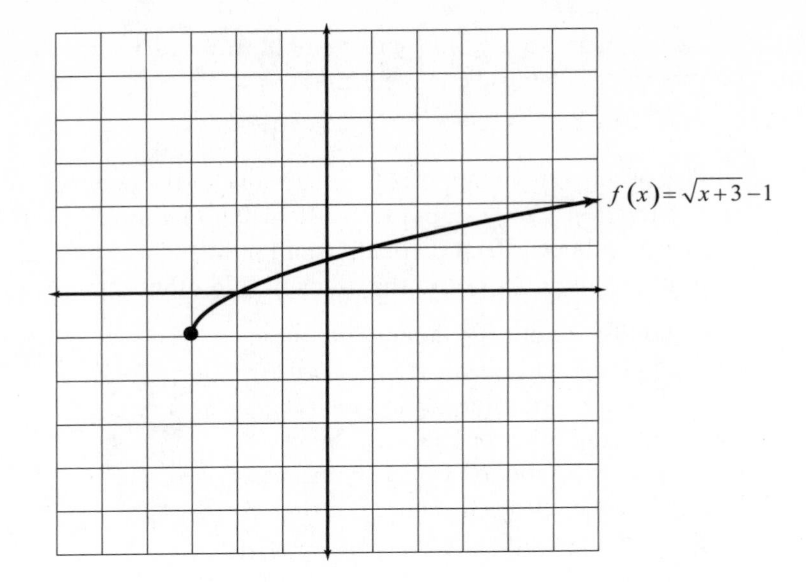

Figure 16.4

The graph of $f(x) = \sqrt{x+3} - 1$.

Solution: Notice that there is no graph to the left of the point (–3,–1). However, from there it continues up and to the right forever and ever. The vertical line $x = -3$ will pass right through that key point, and since it intersects the graph, $x = -3$ belongs to the domain.

Any vertical lines corresponding to numbers less than $x = -3$ (for example, $x = -3.5$, $x = -4$, $x = -5$, and so on) will not intersect the graph, whereas vertical lines based on values greater than $x = -3$ will. Therefore, the domain of $f(x)$ is $x \geq -3$.

Now, imagine horizontal lines all over the coordinate plane to determine the range. The horizontal line $y = -1$ will slice right through the point (–3,–1), but any horizontal lines below it will pass harmlessly beneath the graph. On the other hand, any horizontal line above $y = -1$ will hit the graph somewhere (even if it doesn't do so on this small grid); even though the graph is shallow, it will eventually reach infinite heights. Therefore, the range of the function is $y \geq -1$.

You've Got Problems

Problem 3: Once again, consider the graph of $g(x) = -|x - 5| + 3$ you generated in Problem 1 and examined in Problem 2. Find the domain and range of g.

Important Function Graphs

It's sort of a tease to tell you the sorts of things you can do with a function graph without actually helping you make the graph, too. Of course, you could always plot points to make any graph in the world, but there are quicker ways to get pretty good graph sketches without spending years of your life creating them. However, before I show you how to make those sketches, you'll need to memorize some graphs. (You're basically memorizing the general shapes—remember, these will be quick sketches, not exact graphs you're making.)

In Figure 16.5, you'll find the five basic graphs you'll make in algebra. If you want to see where these graphs come from, you can plot points like you did at the beginning of the chapter. Since these graphs are the building blocks for the sketching method I explain in the next section, it's important that you can create them from memory without always having to generate them from scratch. Even if you hate memorizing things, trust me, it will save you tons of time later.

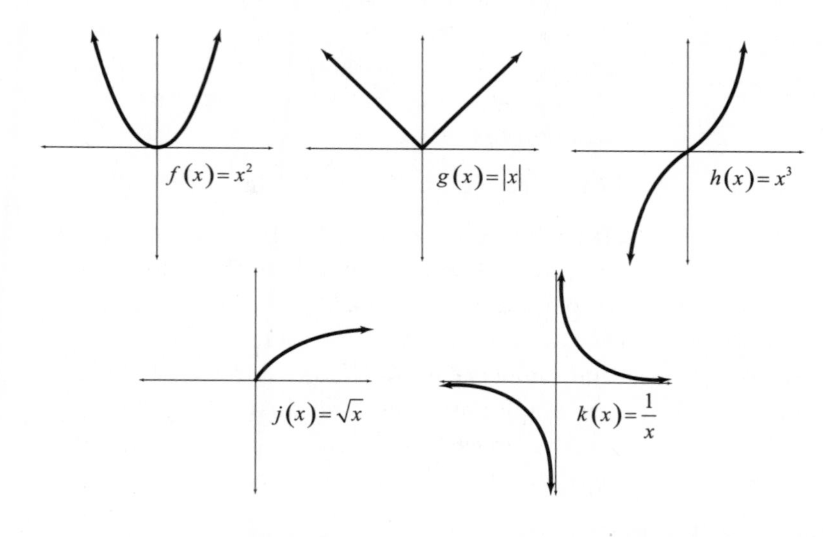

Figure 16.5

The five basic algebra graphs you should know by heart.

Here are brief descriptions of each graph, to help you store them away in the recesses of your brain. Make room in there if you have to. After all, there's no practical use in remembering all those television theme songs you've got memorized; try and forget them to make some room for algebra. (By the way, I used a different letter for each function, but that's just so you can tell them apart; you can, of course, use any letter name for any function.)

- **$f(x) = x^2$ (Domain: all reals; Range: $y \geq 0$)**

  Even without consulting the domain, you would have known that you could square any real number, and the range makes sense, too—the square of a positive or a negative will always be positive. Notice that the graph changes direction at the origin; if you go from left to right, it's decreasing until (0,0) but then increasing after that.

- **$g(x) = |x|$ (Domain: all reals; Range: $y \geq 0$)**

  The absolute value function is another beast that takes any number in but only lets positive numbers out. It, too, has the origin as its key feature, since the graph changes direction there.

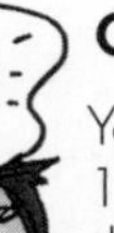

Critical Point

You may notice that Figure 16.5 contains graphs of degree 2 and 3, but not a linear equation with degree 1. That's because you already learned how to graph linear equations in Chapters 5 and 6; just replace the $f(x)$ in a linear function with y, and you can use those old techniques to graph linear functions.

- **$h(x) = x^3$ (Domain: all reals; Range: all reals)**

  The cubic function is not all that exciting. Notice, however, that you can get negative outputs, since a negative input cubed will result in a negative output. The graph's key feature is a little twist at its middle that forces it to pass through the origin.

- **$j(x) = \sqrt{x}$ (Domain: $x \geq 0$; Range: $y \geq 0$)**

  If you're only focusing on real, and not complex numbers, then you can only take the square root of positive things. In addition, a square root always outputs a positive value, so the domain and range are harshly restricted here. In fact, the graph only appears in the first quadrant, because in each of the other quadrants, either x, y, or both x and y are negative.

Critical Point

Since the graph of $k(x)$ gets close to, but never touches, the lines $x = 0$ and $y = 0$, those lines are called the *asymptotes*

- **$k(x) = \frac{1}{x}$ (Domain: $x \neq 0$; Range: $y \neq 0$)**

  Even though the graph gets close to the vertical line $x = 0$ and the horizontal line $y = 0$, it won't touch either one. Plugging 0 into k breaks that "don't divide by zero" rule, so 0 is not in the domain. Furthermore, there's nothing you can plug into k that will give you 0 as an output, so 0 is not in the range, either. This is the only graph of the five that does not contain the origin for those reasons. However, every real number *except* 0 belongs in both the domain and range.

Spend some quality time with these graphs to memorize them. Make sure that, given only an equation like $j(x) = \sqrt{x}$, you could draw the graph without peeking at Figure 16.5. Once you can do that, it's time to put all that boring memorization to good use.

Graphing Function Transformations

Most of the graphs you create in algebra are just funky versions of the graphs in Figure 16.5. In fact, all you have to do to sketch the vast majority of graphs is to take those five basic building blocks and transform them. What do I mean by "transform" the graphs? Simple adjustments, like moving them around the coordinate plane, stretching or squishing them a little, or flipping them around one of the axes. All these things can be accomplished by sticking a number here or a negative sign there—very small changes make a big difference in a graph.

Keep in mind that the sketches you create through function transformations are not exact, but that's okay. If you needed a perfect graph, you could always plot 500 points or use a computer or graphing calculator. In most cases, quick sketches are fine, and besides, learning to graph this way teaches you how all the coefficients and constants in a function affect that function's graph.

Making Functions Flip Out

The first way to transform a function is to make it flip over either the x- or y-axis. Like a lion tamer at the circus who must only blow his whistle to make hungry lions do an incredibly complicated trick (like waltzing with each other to the tune of "Do That to Me One More Time" by Captain and Tennille), you have to expend very little effort to make the graph reflect itself across an axis—just stick a negative sign in the right place.

Specifically, given a function $f(x)$, the graph of:

- **$-f(x)$ is the reflection across the x-axis.** In other words, if you multiply the function itself by -1, the graph will flip over the x-axis; each of its x-coordinates will stay the same, but every corresponding y-coordinate will be its opposite. Therefore, if the graph originally contained the point (a,b), it will now contain $(a,-b)$.
- **$f(-x)$ is the reflection across the y-axis.** If you plug $-x$ into the function, rather than just plain x, it is the x-coordinate that will change to its opposite instead.

Each of these transformations is illustrated in Figure 16.6. In the left graph, the absolute value function is multiplied by -1 and its graph is thus reflected across (flipped over) the x-axis. In the right graph, the radical contains a $-x$ plugged into it, rather than just an x in there, so its graph is reflected across the y-axis.

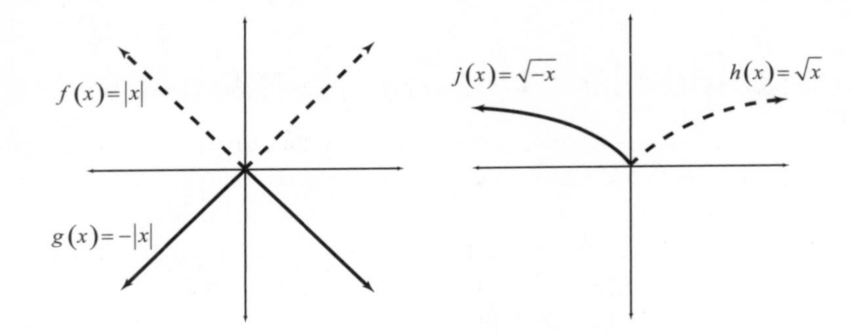

Figure 16.6

The location of the negative sign dictates what sort of reflection will occur.

Stretching Functions

You can use more than negative signs to transform functions; numbers have a profound effect as well. One type of multiplication affects the function's *height*, and the other affects its *width:*

- **The graph of $a \cdot f(x)$ will be a times as tall as $f(x)$.** If you multiply a function by a number, then each of the outputs get multiplied by that number. For instance, each output of some function $4g(x)$ will be exactly four times as high as it was in the original graph of $g(x)$. Of course, if $a < 1$, then the graph will get squished toward the x-axis, rather than stretched vertically. For instance, if $a = \frac{1}{2}$, then each point will only be one half as high.

- **The graph of $g(bx)$ will be $\frac{1}{b}$ times as wide as $g(x)$.** This rule is a little more bizarre. Think of it this way: If you're plugging bx into a function instead of x (where b is a real number), then the graph will get horizontally squished toward the origin by a factor of b if $b > 1$; if $b < 1$, the graph gets horizontally stretched out by a factor of b.

Critical Point

Remember, $\frac{1}{b}$ just means "the reciprocal of b," so that's why the expression $\frac{1}{\frac{1}{3}}$ equals 3, because the reciprocal of $\frac{1}{3}$ is 3.

As you can see in the left graph of Figure 16.7, multiplying $f(x) = \sqrt{x}$ by 2 causes its graph to stretch twice its original height. However, the right graph shows that plugging $\frac{1}{3}x$ into $h(x) = x^2$ causes the graph to stretch out $\frac{1}{\frac{1}{3}}$, or 3, times as wide.

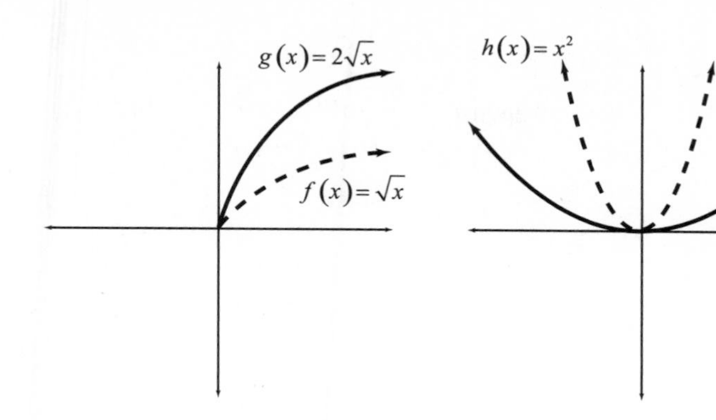

Figure 16.7

Multiplying a function by a number causes the graph to stretch or squish.

Moving Functions Around

The last transformation I'll focus on also involves numbers, but this time, instead of multiplying them, you'll add them to a function.

- **Adding to or subtracting from a function moves its graph up or down, respectively.** So, the graph of $f(x) + 7$ consists of all the points on the graph of $f(x)$ moved up seven units, and the graph of $g(x) - 1$ is just the graph of $g(x)$ moved down one unit.
- **Plugging $x + a$ into a function moves its graph a units left or right.** If $a < 0$, the graph will move right, and if $a > 0$, then the graph moves left a units. This is the opposite of what intuition tells you: Subtracting a actually causes the graph to move *right*, and adding a causes the graph to move left.

In other words, adding to or subtracting from a function moves it vertically, but doing the same thing *inside* a function (adding to or subtracting from the x input) moves it horizontally. Figure 16.8 illustrates these horizontal and vertical graph shifts.

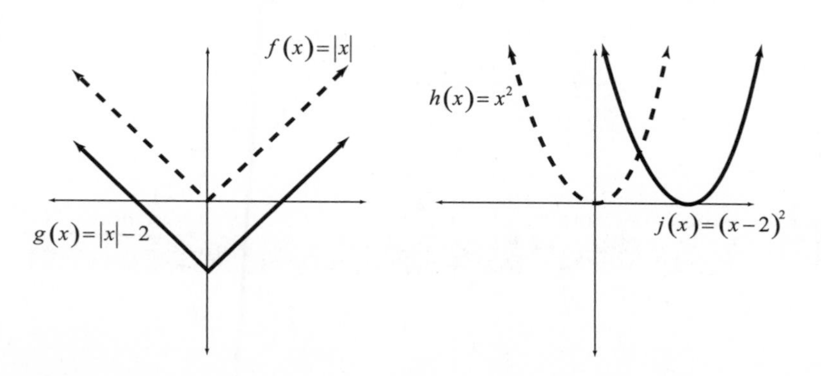

Figure 16.8

You can make a graph jump up, down, right, or left, depending upon where you add or subtract numbers in its equation.

Multiple Transformations

When you need to graph functions that have more than one of the above transformations, you should do them one at a time, in the order you learned them:

1. Reflections
2. Stretching/squishing
3. Vertical/horizontal shifting

See? Graphing functions is as easy as 1-2-3!

Example 3: Graph $f(x) = -2|x+3| + 5$.

Solution: Start with graph of the basic absolute value function, $|x|$. Notice that the version of the absolute value in *f* is multiplied by a –2, so it is flipped across the *x*-axis, and will be stretched vertically by a factor of 2. In addition, the graph will be moved 3 units left and 5 units up. The final graph is shown in Figure 16.9.

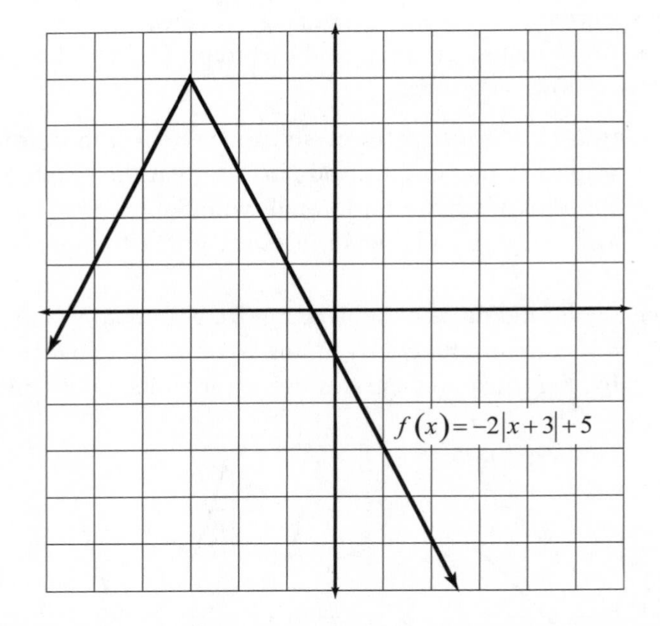

Figure 16.9

The graph of $f(x) = -2|x+3| + 5$ *is just a souped-up version of the graph of* $g(x) = |x|$.

You've Got Problems

Problem 4: Graph $g(x) = (-x)^3 - 2$.

The Least You Need to Know

- The domain of a function is the collection of its valid inputs, and the range of a function consists of all the valid outputs.
- The vertical line test helps you determine whether or not a graph is actually the graph of a function, and the horizontal line test can determine if the function is one-to-one.
- Inserting a negative into a function causes its graph to reflect about one of the axes.
- Multiplying a function by a number will cause its graph to stretch or squish horizontally or vertically.
- Adding to or subtracting from a function causes either a horizontal or vertical shift in its graph.

Please, Be Rational!

Way, way back in the beginning of the book, I spent an entire chapter reviewing fractions with you. You were reintroduced to concepts like least common denominator (along with other terms that make full-grown adults cry) and you reviewed how to do things like add and divide fractions. I wasn't reviewing those things just to upset you. In fact, in this part, the other shoe drops, and you'll work with fractions containing polynomials in their numerators and denominators.

Chapter 17

Rational Expressions

In This Chapter

- Reducing fractions containing polynomials
- Performing basic operations on rational expressions
- Eliminating fractions within fractions

There are lots of fancy words in the English language that describe nasty things; even the most distressing concepts don't sound so bad if they have a pretty name. Politicians know the power that certain words carry and are especially good at avoiding those and choosing other, less controversial terms. For example, who wants to go to "war," when simply being involved in a "conflict" sounds much less threatening? Most people are opposed to "tax increases," but wouldn't be automatically opposed to an "income adjustment."

In mathematics, when you mention the word "fraction," people panic. They sweat, their eyes dart about nervously, and then they sprint from the room. As a former high school math teacher, I know this from experience.

One morning, in a basic algebra class, I announced that we would be discussing fractions, and a student in the front row bolted like a frightened deer. He ran for the door, but in the process, managed to snag his pant leg on another student's desk. Caught in his innate, instinctual flight response, he couldn't think straight to turn around and free his clothing from the obstruction, and, instead, *chewed his own leg off to get free.*

Well, maybe that's a bit of an exaggeration. (He actually just tripped over a desk on the way to sharpen his pencil, but that's not as dramatic a story.) However, whenever I mentioned fractions, students did get wide-eyed, and most got very tense. However, if I said we'd be discussing "rational expressions," it went over much better. Even though you know from Chapter 1 that a rational number (and likewise a rational expression) is just a number that can be expressed as a fraction, for some reason the word "rational" just doesn't have the same pain and fear attached to it as the word "fraction"—just like the word "child entertainer" doesn't inspire the fear that the phrase "scary maniacal clown" does.

Simplifying Rational Expressions

The plan of action for this chapter should be familiar; once again, with the introduction of a new concept, you'll learn how to add, subtract, multiply, and divide the new element. However, since rational expressions are just fractions, one skill precedes learning those operations—you need to be able to reduce those fractions also.

Before I show you how to simplify fractions containing polynomials, though, I want to revisit the process I described for simplifying numeric fractions in Chapter 2. Back then, I told you that the best way to reduce a fraction was to divide both its numerator and denominator by any common factors. Now, I want to show you why that works.

Consider the fraction $\frac{12}{30}$. Rewrite the numerator and denominator as a product of prime factors. This means breaking down each number into a product and then factoring those numbers until all you're left with are prime numbers. For example, here are two numbers that multiply to get 12 (the numerator): $4 \cdot 3$. However, that's not a prime factorization because 4 is a composite number. So, rewrite 4 as a product as well: $2 \cdot 2 \cdot 3$. Since all of those numbers are prime, that's the prime factorization you're looking for.

Critical Point

Even if you had started with $2 \cdot 6$ in your attempt to get the prime factorization of 12, when you further factored the 6, you'd have gotten $2 \cdot 2 \cdot 3$, the same prime factorization. That's because (according to a theorem called the Fundamental Theorem of Algebra) every number has its own, unique prime

Now it's time to factor 30 into prime numbers. Notice that $3 \cdot 10 = 30$, but 10 is not a prime number. However, factor it into $2 \cdot 5$, and you'll get the prime factorization for 30: $2 \cdot 3 \cdot 5$. (I just wrote the factors from least to greatest.) When you've finally got the numerator and denominator factored, rewrite the fraction using those strings of factors.

$$\frac{12}{30} = \frac{2 \cdot 2 \cdot 3}{2 \cdot 3 \cdot 5}$$

Remember, the whole purpose of this exercise was to simplify the fraction $\frac{12}{30}$, and you're moments away from accomplishing that goal. If any factor in the numerator has a twin in the denominator, cross those two twin numbers out. In this example, you can cross out a pair of 2's and a pair of 3's.

$$\frac{\cancel{2}\cdot 2\cdot \cancel{3}}{\cancel{2}\cdot \cancel{3}\cdot 5}$$

The fraction in reduced form is $\frac{2}{5}$.

Basically, this process tells you that simplifying a fraction boils down to completely factoring the numerator and denominator and then eliminating matching pairs of factors on opposite sides of the fraction bar. You'll do the same thing to simplify fractions containing polynomials—the factors will just look a little different, that's all.

How'd You Do That?

You're allowed to cross out matching factors in the numerator and denominator because, technically, anything divided by itself is 1.

$$\frac{\cancel{2}\cdot 2\cdot \cancel{3}}{\cancel{2}\cdot \cancel{3}\cdot 5} = 1\cdot 1\cdot \frac{2}{5}$$

Example 1: Simplify the expression $\dfrac{9-x^2}{x^2+x-12}$.

Solution: Begin by factoring the numerator and denominator; notice that the numerator is the difference of perfect squares.

$$\frac{(3-x)(3+x)}{(x+4)(x-3)}$$

Uh oh—I don't see any matching factors. Even though $(3 - x)$ and $(x - 3)$ are close, they don't match. However, they will match if you factor a –1 out of $(3 - x)$:

$$\frac{-1(-3+x)(3+x)}{(x+4)(x-3)}$$

Think about it: $-3 + x$ and $x - 3$ are really the same thing (thanks to the commutative property for addition—you're just adding an x and a –3 in a different order in each one), so you can cross out those matching factors.

$$\frac{-1\cancel{(-3+x)}(3+x)}{(x+4)\cancel{(x-3)}}$$

Kelley's Cautions

When you get the answer of $\frac{-3-x}{x+4}$ in Example 1, some students are tempted to cross out the x's in the numerator and denominator. You can't do that, because the x's are now being *added;* remember, you can only eliminate pairs of matching factors that are *multiplied,* like you did earlier in the problem.

Multiply what's left in the numerator to get the numerator of the reduced fraction; since only $x + 4$ is left in the denominator, it must be the denominator of the reduced fraction.

$$\frac{-1(3+x)}{x+4}$$

It's common to leave a reduced rational expression in factored form like this (especially later, when you'll have lots of factors left in the numerator and denominator), but it's just as correct to distribute the –1 in the numerator, as well.

$$\frac{-3-x}{x+4}$$

You've Got Problems

Problem 1: Simplify the expression $\frac{2x^3+5x^2-3x}{6x^2+7x-5}$.

Combining Rational Expressions

Since (once the fancy talking has been stripped away) rational expressions are just plain old fractions, adding or subtracting them will still require you to find a least common denominator (LCD). Unfortunately, the Bubba technique I described to calculate the LCD in Chapter 2 doesn't work once variables get tossed into the mix, so you'll need to use a more formal method.

If finding an LCD is like going to a dance, then consider this method your dressy suit pants, while the Bubba technique is a well-worn pair of jeans. Both keep you clothed, but neither is appropriate in all circumstances—it depends upon the kind of dance you're going to. There's nothing wrong with going informal if a fraction contains only numbers, but if you spot variables, you have to get all gussied up with the formal method. Actually, if given the choice, I'd prefer not going at all, since (from afar) my dance technique looks like someone walking into an electrified fence.

To calculate an LCD for a rational function, follow these steps:

1. Factor all denominator polynomials completely.
2. Make a list that contains one copy of each factor, all multiplied together.
3. The power of each factor in that list should be the highest power that factor is raised to in any denominator.
4. The list of factors and powers you generated is the LCD.

Critical Point

You can use this formal method of calculating the LCD with numbers as well—just use prime factorizations of the numbers. For instance, to calculate the LCD of $\frac{5}{36}$ and $\frac{11}{40}$, start by generating the prime factorizations of the denominators.

$$36 = 4 \cdot 9 = 2 \cdot 2 \cdot 3 \cdot 3 = 2^2 \cdot 3^2$$

$$40 = 5 \cdot 8 = 5 \cdot 2 \cdot 2 \cdot 2 = 2^3 \cdot 5$$

There are three different factors: 2, 3, and 5. The LCD will be the product of those factors, if each one is raised to the highest power it achieves in the factorizations. (Since 2 appears twice, as 2^2 and 2^3, you should use the 2^3 version in the LCD since it represents the higher power.)

Therefore, the LCD of those fractions is:

$$2^3 \cdot 3^2 \cdot 5 = 8 \cdot 9 \cdot 5 = 360$$

Once you figure out the least common denominator, you can apply it to the problem (by forcing all the denominators to match) and then add the fractions together.

Example 2: Simplify the expression.

$$\frac{2x+4}{x^2-12x+36} - \frac{3x}{x^2+2x-48}$$

Solution: Begin by factoring the denominators.

$$\frac{2x+4}{(x-6)^2} - \frac{3x}{(x+8)(x-6)}$$

There are two different factors in this problem, $(x - 6)$ and $(x + 8)$, so list them multiplied together: $(x - 6)(x + 8)$; you're not finished yet, however. Notice that $(x - 6)$ appears in both denominators, so when creating the LCD, you should use the one raised to the higher power: $(x - 6)^2$ instead of $(x - 6)^1$. Therefore, the LCD is $(x - 6)^2(x + 8)$. Don't multiply the factors of the LCD together—leave them as is.

Compare each denominator with the LCD. What factor, if any, does each denominator need to match the LCD? The left fraction needs $(x + 8)$, and the right fraction needs another $(x - 6)$. Multiply both the numerator and denominator of each fraction by the factor or factors it needs to make its denominator complete.

$$\frac{(x+8)}{(x+8)}\cdot\frac{2x+4}{(x-6)^2}-\frac{3x}{(x+8)(x-6)}\cdot\frac{(x-6)}{(x-6)}$$

Multiply only the numerators together, leaving the (now-matching) denominators alone.

$$\frac{2x^2+12x+32}{(x-6)^2(x+8)}-\frac{3x^2-18x}{(x-6)^2(x+8)}$$

Now that the fractions have common denominators, you can combine their numerators and write them together over the common denominator. Notice, though, that the second fraction is subtracted, so that preceding negative sign will need to be distributed through its entire numerator.

$$\begin{aligned}&\frac{2x^2+12x+32-\left(3x^2-18x\right)}{(x-6)^2(x+8)}\\&=\frac{2x^2+12x+32-3x^2+18x}{(x-6)^2(x+8)}\\&=\frac{-x^2+30x+32}{(x-6)^2(x+8)}\end{aligned}$$

At this point, you should check to see if the numerator can be factored; if it can, then perhaps the fraction can be simplified. However, in this case, the numerator is prime. Feel free to leave your answer like this, unless your instructor requires you to expand the denominator by multiplying all of the terms together.

The hardest part of this problem is actually forcing the fractions to contain the least common denominator. Once that's finished, it's all downhill.

You've Got Problems

Problem 2: Simplify the expression $\frac{x}{x^2-4}+\frac{8}{x^2-7x+10}$.

Multiplying and Dividing Rationally

My nine-month-old son Nicholas has been eating lumpy food for a couple of weeks now; he's not up to solid food yet, and can't actually feed himself, but if you spoon anything mashed into his maw, he will slurp it down. Sometimes, that is. There's always the chance he'll sneeze with no warning, and spray my face and glasses with half-chewed green bean casserole (which is a real treat if I happened to have my mouth open at the time).

He's also good at the spray canlike tongue raspberry, or the less-fancy move in which he just spits whatever's in his mouth right onto his lap. Of course, his digestive system works the same as mine, and once he decides to eat the food in his mouth, the mechanics work the same way they do with me—chew, swallow, digest, create foul-smelling diaper. (Actually, the final step there is all his own.)

Basically he and I eat and digest the same way, but when he eats, things are just a lot messier. (I rarely get food in my eyebrows, for instance, especially since I've turned 30.) Similarly, multiplying and dividing with rational expressions works just like multiplying and dividing fractions, only it's sloppier with the larger, variable-heavy expressions.

Remember, you don't need common denominators to calculate a product or a quotient. Multiplying fractions couldn't be easier—you just multiply the numerators together and write the answer over the product of all the denominators:

$$\frac{a}{b}\cdot\frac{c}{d}=\frac{ac}{bd}$$

Don't forget that division is really the same as multiplying by a reciprocal. Just flip the second fraction upside down and make the division symbol into a multiplication sign:

$$\frac{a}{b}\div\frac{c}{d}=\frac{a}{b}\cdot\frac{d}{c}=\frac{ad}{bc}$$

Example 3: Simplify the expressions.

(a) $\frac{2x-1}{x^2-2x-15}\cdot\frac{x-5}{2x^2-5x-3}$

Solution: Rewrite this expression as one fraction, keeping the numerators on top and the denominators on the bottom. However, instead of actually multiplying them together with the distributive property, factor everything.

Kelley's Cautions

Don't forget to check and see if your answer can be simplified, once you're finished multiplying and dividing.

$$\frac{(2x-1)(x-5)}{\left(x^2-2x-15\right)\left(2x^2-5x-3\right)}$$
$$=\frac{(2x-1)(x-5)}{(x-5)(x+3)(2x-1)(x+3)}$$

Kelley's Cautions

If all the factors in either the numerator or the denominator are eliminated when simplifying, place a 1 where the factors used to be, like in the final step of Example 3(a). Some students are tempted to put a 0 there instead, because they see nothing left, but that is incorrect.

Simplify the expression by crossing out pairs of matching factors in the numerator and denominator.

$$\frac{\cancel{(2x-1)}\cancel{(x-5)}}{\cancel{(x-5)}(x+3)\cancel{(2x-1)}(x+3)}$$
$$=\frac{1}{(x+3)^2}\text{ or }\frac{1}{x^2+6x+9}$$

Either of those forms of the answer is correct.

(b) $\frac{x^4-4x^3-21x^2}{4x^2-9}\div\frac{2x^3-19x^2+35x}{8x^3+27}$

Solution: Start by taking the reciprocal of the right-hand fraction and changing this from a division to a multiplication problem. Notice that the left-hand fraction remains unchanged.

$$\frac{x^4-4x^3-21x^2}{4x^2-9}\cdot\frac{8x^3+27}{2x^3-19x^2+35x}$$

Write the product as a single fraction, factoring every one of those expressions. (You'll have to factor the sum of perfect cubes, so review the formula in Chapter 11 if necessary.) To make simplifying easier, I'll rewrite the monomial factor x^2 in its factored form, $x \cdot x$. (One of those x's will cancel out with the x in the denominator.)

$$\frac{\left(x^4-4x^3-21x^2\right)\left(8x^3+27\right)}{\left(4x^2-9\right)\left(2x^3-19x^2+35x\right)}$$
$$=\frac{x\cdot x(x-7)(x+3)(2x+3)\left(4x^2-6x+9\right)}{(2x+3)(2x-3)(x)(x-7)(2x-5)}$$

Reduce the fraction.

$$\frac{\cancel{x}\cdot x\cancel{(x-7)}(x+3)\cancel{(2x+3)}\left(4x^2-6x+9\right)}{\cancel{(2x+3)}(2x-3)\cancel{(x)}\cancel{(x-7)}(2x-5)}$$
$$=\frac{x(x+3)\left(4x^2-6x+9\right)}{(2x-3)(2x-5)}$$

How'd You Do That?

Remember exponential Rule 2 from Chapter 3? It said that $\frac{x^a}{x^b} = x^{a-b}$. Example 3(b) demonstrates why this is true. In that example, I rewrote x^2 as $x \cdot x$, an equivalent repeated multiplication expression, and then canceled out factors to reduce the fraction.

Consider the fraction $\frac{x^{10}}{x^7}$; according to the rule, that should equal x^{10-7}, or x^3. If you write out the repeated multiplication, you can see why that's true:

$$\frac{\cancel{x} \cdot \cancel{x} \cdot \cancel{x} \cdot \cancel{x} \cdot \cancel{x} \cdot \cancel{x} \cdot \cancel{x} \cdot x \cdot x \cdot x}{\cancel{x} \cdot \cancel{x} \cdot \cancel{x} \cdot \cancel{x} \cdot \cancel{x} \cdot \cancel{x} \cdot \cancel{x}}$$

Seven of the x's in the numerator cancel with the seven x's in the denominator, leaving behind $\frac{x^3}{1}$.

Once again, there's no need to multiply those factors together; a final answer in factored form is just dandy.

You've Got Problems

Problem 3: Simplify the expression $\frac{x^2 - 4x - 12}{3x^2 - 10x - 8} \div \frac{x^2 - 3x - 18}{3x + 2}$.

Encountering Complex Fractions

If you hate fractions, then you'll be no fan of *complex fractions*. Just the name alone sounds scary, right? Fractions are hard enough, but *complex* fractions? Great! I imagine that brain surgery is pretty hard to do, but *complex* brain surgery sounds even worse. Actually, your gut fear is probably unjustified because the term "complex fraction" is false advertising for two reasons:

- The word "complex" might suggest that the fractions contain complex numbers, but they don't.
- You already know how to work with complex fractions; you just don't know that I know that you know how to. (But you know now.)

Enough mystery—let's cut to the chase. A *complex fraction* is a fraction that contains a fraction in its numerator or denominator (or both). Complex fractions are considered bad form, so your final answers shouldn't contain them; however, since a fraction

> **Talk the Talk**
>
> A **complex fraction** is a fraction that contains other fractions in its numerator, denominator, or both. It's sort of a double-decker fraction. (Complex fractions are also called *compound fractions*.)

translates into a division problem, they are extremely easy to simplify—just divide the numerator by the denominator.

Example 4: Simplify the complex fraction.

$$\frac{\frac{3x}{x-2}}{\frac{9x^2}{7x-14}}$$

Solution: Rewrite the complex fraction as a quotient—the top fraction divided by the bottom fraction.

$$\frac{3x}{x-2} \div \frac{9x^2}{7x-14}$$

Suddenly, this has become a division problem very similar to Example 3(b). A little multiplication by the reciprocal should do the trick. Don't forget to factor.

$$\frac{3x}{x-2} \cdot \frac{7x-14}{9x^2}$$
$$= \frac{3 \cdot x \cdot 7 \cdot (x-2)}{(x-2) \cdot 9 \cdot x \cdot x}$$

Just simplify the fraction and you're done!

$$\frac{3 \cdot \cancel{x} \cdot 7 \cdot \cancel{(x-2)}}{\cancel{(x-2)} \cdot 9 \cdot \cancel{x} \cdot x}$$
$$= \frac{3 \cdot 7}{9 \cdot x}$$
$$= \frac{21}{9x}$$

You're not quite done yet—you can still simplify the fraction further, since 21 and 9 are both divisible by 3.

$$\frac{7}{3x}$$

You've Got Problems

Problem 4: Simplify the complex fraction $\dfrac{\dfrac{x^2-7x+12}{x+2}}{\dfrac{x-3}{x^2+4x+4}}$.

The Least You Need to Know

- To simplify rational expressions, factor the numerator and denominator and cancel out pairs of factors that appear in each.
- Rational expressions must have common denominators in order to add or subtract them.
- Change rational division problems into multiplication problems by taking the reciprocal of the second fraction.
- Complex fractions are really just quotients in disguise.

Chapter 18

Rational Equations and Inequalities

In This Chapter

- Solving equations containing fractions
- Using cross multiplication to simplify proportions
- Exploring direct and indirect variation
- Solving rational inequalities

Knowing how to add or multiply fractions is all well and good, but what happens when they start popping up in other places? What do you do when an equation contains fractions? How in the world do you solve an inequality featuring fractions? How are you supposed to react if a fraction sneaks into your house when you're asleep and tries to steal that jar you dump your pocket change into? All these questions (with the exception of the last) are answered in this chapter.

Solving Rational Equations

Have you ever noticed that many mathematicians don't have great social skills? (If you haven't noticed, it must be because you don't know many mathematicians.) There's an old joke that goes "How can you spot an extroverted mathematician? He talks to *your* shoes instead of his."

Why is it that some math people can't deal with the outside world? Is it the bright daylight that sears their pupils behind their thick glasses, or the wedgies they endured in school? I don't think either is to blame. I think it's because they're used to a world completely and totally under their control. You see, in the math world, if you don't like something, you can manipulate the rules of the universe and make any unpleasantness disappear.

Case in point: Most mathematicians dislike fractions. It's not because they don't understand them, it's just that the constant need of common denominators is annoying. Therefore, whenever the opportunity arises, math people will completely eliminate fractions from their landscape. For instance, you can very easily eliminate every fraction from an equation just by multiplying everything in that equation by its least common denominator.

CAUTION

Kelley's Cautions

If you multiply an equation by something containing an x, you may be adding incorrect solutions. Always check to make sure any answer you get can be plugged into the original equation. (In other words, make sure none of the denominators become 0 when you plug your answers in for x.)

Example 1: Solve the equation.

$$\frac{1}{x+5}+\frac{x}{x+2}=\frac{2x-1}{x^2+7x+10}$$

Solution: This equation contains three rational expressions; your goal will be to eliminate all of those fractions to make the equation much simpler to solve. Start by factoring any expression you can in the equation. (In this case, the quadratic can be factored.)

$$\frac{1}{x+5}+\frac{x}{x+2}=\frac{2x-1}{(x+5)(x+2)}$$

The least common denominator of all three fractions is $(x + 5)(x + 2)$. If you multiply the *entire equation* by that expression, the fractions will disappear. (Technically speaking, you'll multiply the equation by $\frac{(x+5)(x+2)}{1}$, which is the exact same expression—it just reminds you to multiply the least common denominator by each individual fraction's numerator while leaving the denominators unchanged.)

$$\frac{(x+5)(x+2)}{1}\left[\frac{1}{x+5}+\frac{x}{x+2}=\frac{2x-1}{(x+5)(x+2)}\right]$$

$$\frac{(x+5)(x+2)}{(x+5)}+\frac{(x+5)(x+2)x}{(x+2)}=\frac{(x+5)(x+2)(2x-1)}{(x+5)(x+2)}$$

You can simplify all of those fractions.

$$\frac{\cancel{(x+5)}(x+2)}{\cancel{(x+5)}}+\frac{(x+5)\cancel{(x+2)}x}{\cancel{(x+2)}}=\frac{\cancel{(x+5)}\cancel{(x+2)}(2x-1)}{\cancel{(x+5)}\cancel{(x+2)}}$$

Notice that each of those denominators have been eliminated, and are technically now all equal to 1. However, there's no need to write a denominator of 1, so you can rewrite the equation using only the numerators.

$$(x + 2) + (x + 5)x = 2x - 1$$

Distribute the x in the second term and combine all like terms (by setting the equation equal to 0).

$$x + 2 + x^2 + 5x = 2x - 1$$

$$x^2 + 6x + 2 = 2x - 1$$

$$x^2 + 4x + 3 = 0$$

Hey! There's a plain old quadratic equation left over that you can solve by factoring.

$$(x + 3)(x + 1) = 0$$

$$x = -3 \text{ or } x = -1$$

If you plug both of those solutions into the original equation, you get true statements, so they are both valid answers.

Check $x = -3$

$$\frac{1}{-3+5}+\frac{-3}{-3+2}=\frac{2(-3)-1}{(-3)^2+7(-3)+10}$$

$$\frac{1}{2}+\frac{-3}{-1}=\frac{-7}{-2}$$

$$\frac{1}{2}+3=\frac{7}{2} \quad \underline{\text{True}}$$

Check $x = -1$

$$\frac{1}{-1+5}+\frac{-1}{-1+2}=\frac{2(-1)-1}{(-1)^2+7(-1)+10}$$

$$\frac{1}{4}+\frac{-1}{1}=\frac{-3}{4}$$

$$\frac{1}{4}-1=-\frac{3}{4} \quad \underline{\text{True}}$$

You've Got Problems

Problem 1: Solve the equation $\frac{x+3}{x-8}+\frac{x}{x^2-6x-16}=1$.

Proportions and Cross Multiplying

Of all the rational equations you'll see as an algebra student, one of the most common will be the *proportion*, which is an equation in which two fractions are set equal to one another, like this:

$$\frac{a}{b}=\frac{c}{d}$$

Talk the Talk

An equation that sets two fractions equal is known as a **proportion;** such equations can be solved by means of **cross multiplication,** in which you multiply the numerator of one fraction by the denominator of the other and set those products equal.

You can solve a proportion just like any other rational function, if you multiply it by the least common denominator of the fractions and simplify. However, there's a quicker and cooler shortcut called *cross multiplication* that works for proportions only. All you do is multiply each numerator by the denominator of the other fraction and set the results equal, as shown in Figure 18.1.

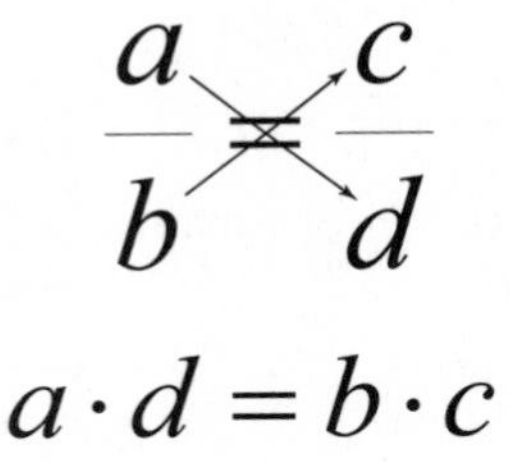

$$\frac{a}{b}=\frac{c}{d}$$

$$a\cdot d=b\cdot c$$

Figure 18.1

Cross multiplying eliminates the denominators in a proportion quickly, without the need to calculate a least common denominator.

Example 2: Solve the equation.

$$\frac{x+3}{x-1}=\frac{x+6}{2}$$

Solution: Since this is a proportion, you can cross multiply to eliminate the fractions.

$$(x+3)\cdot 2=(x-1)(x+6)$$

Distribute the 2 on the left side and multiply the binomials on the right.

$$2x+6=x^2+5x-6$$

This is a quadratic equation, so set it equal to 0 to try and solve it by factoring.

$$x^2+3x-12=0$$

Nuts—this bad boy isn't factorable, so you'll have to resort either to completing the square or the quadratic formula to get the solution. (I usually use the quadratic formula, so it doesn't feel like I memorized it for nothing.) You'll end up with a solution of:

$$x = \frac{-3 \pm \sqrt{57}}{2}$$

Okay, so that wasn't the most attractive problem in the world, but did you notice that the tough part was solving the quadratic equation, not dealing with the fractions? Here's even better news: You don't have to test these (horribly disgusting and unforgivably irrational) answers! Unlike the method you learned earlier in the chapter, cross multiplication won't introduce possibly false answers that require checking! Huzzah!

You've Got Problems

Problem 2: Solve the equation $\frac{3x-2}{x} = \frac{2x}{x+1}$.

Investigating Variation

There is an undeniable cause-and-effect link between many things in life. You may intuitively feel these connections every once in a while and accept their universal truths without even noticing it. In some cases, an increase in one event causes an increase in another; for example, the more french fries you eat at a fast food restaurant, the bigger your waist will expand. The faster you drive, the more likely you are to get a speeding ticket.

On the other hand, there are also events that are paired up in the opposite (or inverse) way: An increase in one leads to a decrease in the other. I was once told that the number of keys you carry around with you at the office is inversely related to how successful you are at work. (In other words, the big boss only has one or two keys, because they open everything, but the worker bees have key rings that might as well belong to a zookeeper.) For those of you calloused to the world of government, how about this inverse relationship: The more successful you are at politics, the less honest you probably are as a person. (Ooh, a cheap shot!)

These relationships are called *variation*, and in this section, you'll learn to express them mathematically using rational equations.

Direct Variation

Direct variation is the relationship I described in which an increase in one value corresponds with an increase in the other, like in the statement "The longer you study for a test, the higher your score will be." In this case, an increase in study hours should correlate with an increase in the percentage test score.

Talk the Talk

If the variables x and y exhibit **direct variation,** then y is exactly k times as large as x; k is called the **constant of proportionality.** Direct variation is also called **proportional variation.**

In algebra, however, you'll only look at a very specific kind of direct variation: When one value is multiplied by a number k (called the *constant of proportionality*), then the corresponding value will increase by exactly k times as well. Mathematically speaking, if x and y are directly related, then

$$y = k \cdot x$$

meaning that y is exactly k times as big as x. For example, if my test score *directly* (or *proportionally*) *varies* with the number of hours I spent studying, then studying twice as long must double my test score. Furthermore, the test score I get after studying for three hours must be exactly three times as large as the test score I would have gotten had I studied for only one hour.

Usually, your first job in a direct variation problem is calculating that constant of proportionality, and you do that by dividing the two numbers which are directly related. So, if y varies directly with x, then

$$\frac{y}{x} = k$$

Once you find k, you can use that equation to find any missing value.

Example 3: Assume that the measurements of the Statue of Liberty's facial features vary directly with the measurements of my facial features. The length of the statue's nose measures 54 inches, and the length of my nose measures 2.375 inches, as illustrated in Figure 18.2. If Liberty's right eye measures 30 inches across, calculate the distance across my right eye. (Round all calculations to the thousandths place.)

Figure 18.2

The Statue of Liberty may have a large nose, but my nose is no small matter either, and I have no pointy hat to draw attention away from it.

Solution: Since direct variation is exhibited, divide the statue's nose measurement by mine to determine the constant of proportionality. I suggest using a calculator unless you're really excited about long dividing by hand.

$$k = \frac{54}{2.375} \approx 22.737$$

Now that you know the value of *k*, it's time to calculate eye width. You should use the equation

$$\frac{y}{x} = k$$

Where do you plug in her eye measurement, for *y* or for *x*? It depends upon where you put her measurement when you calculated *k*. Notice that the 54-inch honker appeared in the numerator, so her enormous measurement must once again appear in the numerator of this equation.

$$\frac{30}{x} = 22.737$$

You can rewrite this equation as a proportion, if you express 22.737 as the equivalent fraction $\frac{22.737}{1}$ and solve via cross multiplication.

$$\begin{aligned} \frac{30}{x} &= \frac{22.737}{1} \\ 30 \cdot 1 &= x(22.737) \\ \frac{30}{22.737} &= x \\ 1.319 &= x \end{aligned}$$

My eye is approximately 1.32 inches across. Of course, it would have just been easier to measure my eye than do all that math, but then again, you couldn't have measured my eye, because you don't have ready access to it. I'm not about to start loaning it to people just so they can get answers quicker. How could I be sure you'd mail it back to me?

You've Got Problems

Problem 3: The variables *x* and *y* vary proportionally, and when $x = 12$, $y = 15$. Determine the value of *y* when $x = 35$.

Indirect Variation

Indirect (or *inverse*) *variation* occurs when an increase in one quantity leads to a decrease in the other. Specifically, when one quantity increases to n times its original amount, the other will decrease to $\frac{1}{n}$ times its original amount.

Talk the Talk

If two quantities, x and y, exhibit **indirect** (or **inverse**) **variation,** then their product remains constant even as the values of x and y change: $xy = k$. This means that as one quantity becomes n times as large, the other must become $\frac{1}{n}$ times as big.

Because of this, the product of the indirectly related quantities will always be a constant:

$$xy = k$$

How is this possible? Watch what happens if x in this equation becomes n times as large and y becomes $\frac{1}{n}$ times as big:

$$(x \cdot n)\left(y \cdot \frac{1}{n}\right) = k$$

$$x \cdot y \cdot n \cdot \frac{1}{n} = k$$

$$xy \cdot 1 = k$$

$$xy = k$$

Critical Point

In *The Simpsons* episode "HOMR," the reason for Homer's lack of intellectual acuity is revealed (in other words, we find out why he's such a dummy): There's a crayon stuck in his brain that's been lodged there since childhood. Doctors remove it, and suddenly, he becomes a brainiac. However, it turns out to be a mixed blessing, because people shun the new, smarter Homer.

Grieved, he goes to talk to his superintelligent daughter, Lisa, who affirms that his worst fears have been realized. "Dad, as intelligence goes up, happiness often goes down. In fact, I made a graph," she says, and holds up Figure 18.3. "I make a lot of graphs," she sighs sadly.

This is the graph of an inverse variation. Notice that as x gets bigger and hence you get smarter, y gets smaller (the graph gets lower and lower the further right you go), meaning your happiness is decreasing.

Inversely, the smaller your intelligence (the closer to the origin you are along the x-axis), the higher the graph stretches, meaning a happier person.

Who says you couldn't learn something from watching television? In fact, *The Simpsons* series features tons of mathematical jokes and references. Check out Dr. Sarah J. Greenwald and Dr. Andrew Nestler's website at www.simpsonsmath.com.

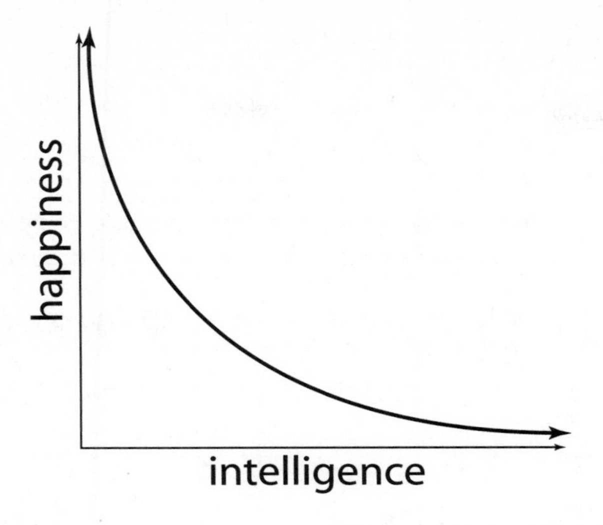

Figure 18.3

Lisa has bad news for her newly intelligent father.

Direct and inverse variation are actually very closely related. Remember, in direct variation, it was the *quotient* of the two values that remained constant $\left(\frac{y}{x} = k\right)$, whereas inverse variation features a constant *product* ($xy = k$). Once again, your first job in an inverse variation problem will be to calculate that constant and then use it in the equation $xy = k$ to calculate missing values.

Example 4: Suppose that the number of times Oscar brushes his teeth in a year varies inversely with the total number of cavities he'll get that year. In 2003, he brushed his teeth a total of 48 times and got four cavities. If he plans to brush his teeth 140 times this year, how many cavities should he expect? (Round your answer to the nearest whole number.)

Solution: Let t equal the total number of times Oscar brushes his teeth in a year and c equal his total number of cavities for that year. Since the problem tells you that t and c vary inversely, you know that their product will be a constant, k:

$$t \cdot c = k$$

In the year 2003, $t = 48$ and $c = 4$, so plug those into the equation to find k:

$$48 \cdot 4 = k$$

$$192 = k$$

This year, t will equal 140, and you have to calculate the corresponding c, knowing that k will still equal 192.

$$t \cdot c = k$$

$$140 \cdot c = 192$$

To solve for c, just divide both sides by 140.

$$c = \frac{192}{140}$$
$$c \approx 1.371$$

The directions indicate that you should round your answer to the nearest whole number, so Oscar should expect one cavity this year, because $c = 1$ approximately.

You've Got Problems

Problem 4: Assume x varies inversely with y, and that $x = 9$ when $y = 75$. Find the value of x when $y = 2$.

Solving Rational Inequalities

At the end of Chapter 13, I showed you how to solve and graph one-variable quadratic inequalities. Do you remember the process? Basically, you factored the quadratic, found critical numbers, split the number line into intervals based on those critical numbers, and then tested those intervals to see which were solutions of the inequality.

You'll use a very similar process to solve rational inequalities. Actually, to be perfectly honest, the process is *exactly* the same—I just didn't give you a completely accurate definition of what a critical number was back then. It's not that I didn't want to tell you; I've just found that algebra students work best on a "need to know" basis. You really didn't need to know the full definition in Chapter 13, but to solve rational inequalities, you do.

Talk the Talk

A **critical number** either causes a function to equal 0 or makes it undefined.

I already mentioned that a *critical number* is an x value that makes an expression equal 0. However, there's another way that values earn the classification of "critical number," when they cause the function to be undefined.

If you follow these steps, solving rational inequalities is a piece of cake:

1. **Rearrange the inequality so that only 0 remains on the right side.** This means you should add and subtract terms from both sides to get everything moved to the left.
2. **Create one fraction on the left side, if necessary.** Use common denominators to combine any terms into one single fraction.

3. **Factor the numerator and denominator.** This makes finding critical numbers extremely easy.
4. **Set each factor in the numerator equal to 0 and solve.** Mark these critical numbers on the number line using an open dot (if the inequality symbol is $<$ or $>$) or a closed dot (if the inequality symbol is $\le$ or $\ge$).
5. **Set each factor in the denominator equal to 0 and solve.** These, too, are critical numbers, but should always be marked with open dot on the number line, since they represent the values that cause the rational function to be undefined. (Remember, a 0 in the denominator is bad news, since dividing by 0 is illegal.)
6. **Choose test points to find solution intervals.** Once you've found the critical numbers, the process is identical to the steps you followed in Chapter 13.

Be extra careful with the dots you place on the number line. If you use the wrong dot, you'll get the inequality signs in your final answer wrong, not to mention that the graph will be inaccurate as well.

Example 5: Solve the inequality and graph its solution.

$$\frac{2x+5}{x+4} \ge -x$$

Solution: Start by moving that $-x$ to the left side of the inequality by adding x to both sides—the right side has to be completely clear of any terms except 0.

$$\frac{2x+5}{x+4} + x \ge 0$$

Your goal now is to create only one fraction on the left side of the inequality by adding everything together. The least common denominator of those two terms is $x + 4$, so multiply the newly relocated $\frac{x}{1}$ term's numerator and denominator by that value and combine the fractions.

$$\frac{2x+5}{x+4} + \frac{x}{1} \cdot \frac{x+4}{x+4} \ge 0$$

$$\frac{2x+5}{x+4} + \frac{x^2+4x}{x+4} \ge 0$$

$$\frac{x^2+6x+5}{x+4} \ge 0$$

Factor the numerator.

$$\frac{(x+1)(x+5)}{x+4} \geq 0$$

Set each factor of the numerator equal to 0 and solve to get two critical numbers.

$$x+1=0 \quad \text{or} \quad x+5=0$$
$$x=-1 \quad \text{or} \quad x=-5$$

You should mark both $x = -1$ and $x = -5$ on the number line with *solid* dots, since the inequality sign $\geq$ allows for equality, as shown in Figure 18.4. The final critical number is generated by setting the denominator's factor equal to 0.

$$x + 4 = 0$$
$$x = -4$$

Remember to use an open dot for $x = -4$, since its value comes from the denominator.

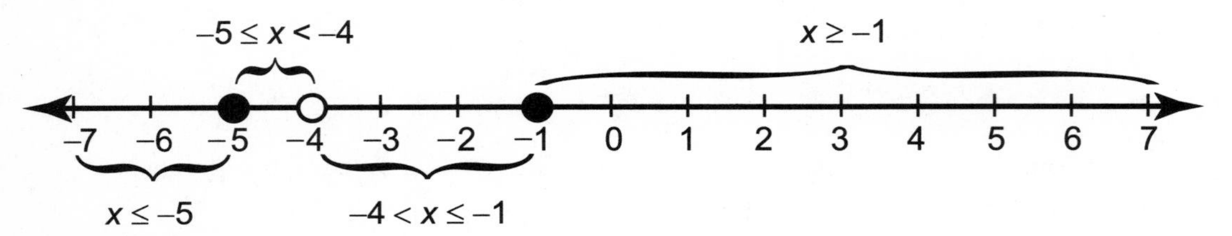

Figure 18.4

The critical points, appropriately marked on the number line.

Now choose test points from each interval (I suggest $x = -6$, $x = -4.5$, $x = -2$, and $x = 0$). Both $x = -4.5$ and $x = 0$ make the inequality true, so their intervals make up the solution: $-5 \leq x < -4$ or $x \geq -1$. Darken those intervals on the number line to make the graph, shown in Figure 18.5

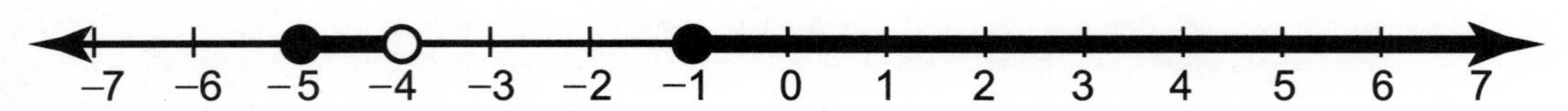

Figure 18.5

The graph of the solution to $\frac{2x+5}{x+4} \geq -x$.

You've Got Problems

Problem 5: Solve and graph the inequality $\frac{x+7}{x-2} < 3$.

The Least You Need to Know

- Eliminate fractions in rational equations by multiplying everything by the least common denominator or by cross multiplying proportions.
- If x varies directly with y, then $y = kx$, but if x varies inversely with y, then $xy = k$ (where k is a real number).
- Critical numbers represent values where a function either equals 0 or is undefined.

Part 7
Wrapping Things Up

You can't write an algebra book without a chapter on word problems. That would be like a dentist visit without getting your teeth drilled, or an appointment with the ophthalmologist that didn't include getting that fabulous burst of air shot straight into your eyeball during the glaucoma test. Word problems are a necessary evil of algebra, jammed in there to show you that you can use algebra in "real life." Once you persevere through the word problems, this part also contains a comprehensive review of all the concepts of algebra, so you can practice to your heart's content.

Chapter 19

Whipping Word Problems

In This Chapter

- Calculating simple and compound interest
- Solving geometric measurement problems
- Determining distance and rate of travel
- Cracking problems with combinations and mixtures

The mythological Greek gods were very creative when dealing out punishments. Take, for example, the extraordinarily unpleasant fate of Sisyphus. The guy was no peach, known as a cruel and heartless king, and was too sly for his own good. When Death itself, in the form of Hades, came to claim him, he managed to shackle Death and hold it captive. Eventually, people started to notice that no one was dying ("Hey, Gary, sorry about running you over with my chariot this morning—I was changing radio stations; no offense, but you don't look so good decapitated") and Sisyphus's luck ran out.

As punishment, he was doomed to roll a large boulder up a steep mountain in the land of the dead for all eternity. That's pretty nasty, but it's not the worst part. After hours and hours of hard labor, moving the mammoth rock slowly, gaining ground at an agonizingly slow rate, and straining every muscle in his body, just before reaching the zenith of the cliff and accomplishing his task, the boulder would roll all the way back down the hill.

I can't imagine such a depressing fate, to spend all of eternity forced to do something that is inherently painful and pointless, knowing that no matter how hard you try, you'll never be able to accomplish your task. However, thousands of students every day engage in their daily battle as Sisyphus. Each morning, they walk up to the giant cliff that is algebra and a massive boulder representing word problems. In fact, their fate is arguably worse.

Even though they are not doomed to work word problems unsuccessfully for all eternity, they lack one thing Sisyphus doesn't—direction. At least he knows *what he's supposed to do!* I have seen lots of students who stare open-mouthed at word problems with great, gleaming eyes, wet with unshed tears, who all say the same thing: "I don't even know where to start! How am I supposed to do these problems if I can't even figure out the first step?" (Of course, there's more cursing when *they* say it, but you get the idea.)

In this chapter, I'm going to risk incurring the wrath of the mythological algebraic deities and free you from your doom. I'll introduce you to four of the most common types of word problems and provide a plan of attack for each one. That way, instead of living in fear, you can stare those problems down and calmly retort (in the voice and comedic timing of a big-screen action hero), "Let's rock." You'll feel much boulder. (Horrible puns definitely intended.)

Interest Problems

There are three good reasons to deposit your life savings in a bank account, rather than hide it in your closet or mattress:

1. A bank is safer, and if your money is stolen, there are usually federal laws that insure your investment.
2. A bank affords you the unique opportunity to write with pens chained to desks. Even though your money allows banks, themselves, to rake in the dough, for some reason they are very adamant that you not accidentally take their pens, each worth only pennies.
3. You earn interest on your money without having to exert any effort at all.

Talk the Talk

The amount of money you initially deposit in an interest problem is called the **principal**.

Interest is a great thing. It's free money that you earn just by keeping your money in a safe place. In algebra, you may be asked to solve problems in which you calculate the interest earned by some initial investment (which is called the *principal*) over some length of time. Specifically, there are two major types of interest problems you may be asked to solve: simple interest and compound interest.

Simple Interest

If your money grows according to simple interest, you're basically just earning a small percentage of your initial investment each year as interest. For instance, if the principal of an account is $100 and your annual interest rate is 6.75%, at the conclusion of every year you will have earned an additional $6.75 (since $6.75 is 6.75% of $100).

Here's the bad news: Even though your account will grow slightly every year, you won't earn more interest! In simple interest problems, you only earn interest on the initial investment, no matter how long you've had an active bank account or how much interest that money has accrued.

The formula for calculating simple interest is:

$$i = prt$$

where p is your principal, r is the annual interest rate expressed as a decimal, and i is the interest you have earned after the money has been invested for t years.

Critical Point

To convert a percent into a decimal, drop the percent sign and multiply by .01. For instance, the decimal equivalent of 6.75% is (6.75)(.01) = .0675. (Conversely, to change a decimal into a percent, multiply by 100 and stick a percent sign on the end. Therefore, the percent equivalent of .45 is (.45)(100) = 45%.)

Example 1: You were a very thrifty and money-savvy child. Instead of spending the money the tooth fairy gave you for your baby teeth, you invested that cash in one lump sum of $32.00 as a teenager, in a bank account with a fixed annual interest rate of 7.75%. What is the balance of the account exactly 30 years later?

Solution: To calculate the balance of an account, just add the interest you earned to the principal. Of course, you still need to figure what that interest is. Use the formula $i = prt$, where $p = 32$, $r = .0775$ (the decimal equivalent of 7.75%), and $t = 30$.

$$i = prt$$

$$i = (32)(.0775)(30)$$

$$i = 74.4$$

You earned $74.40 in interest over that 30-year period, so if you add in the initial investment, your total balance is:

$$\text{balance} = \text{principal} + \text{interest earned}$$

$$= \$32 + \$74.40$$

$$= \$106.40$$

Compound Interest

Most banks don't use simple interest; the more money you deposit, the more money they can potentially make, so they want to encourage you to deposit as much as possible into your account. One way they do this is via *compound interest*, in which you earn money based on your original principal *and* the interest you've accrued to that point.

Let's say you deposit \$100 in an account whose interest is compounded annually at a rate of 6.0%. At the end of the first year, you will have a balance of \$106, just like you would with simple interest. However, at the end of the second year, you'll earn 6.0% interest on the new balance of \$106, not just the original balance of \$100.

Talk the Talk

If your bank account accrues **compound interest,** then you earn interest based upon your entire balance, rather than just the initial investment.

Even better, most banks don't just compound their interest once a year. Whether they compound weekly (52 times a year), monthly (12 times a year), or quarterly (4 times a year) can make a big difference in your bottom line.

Critical Point

The more times the interest in your account is compounded, the more money you'll earn. The best possible scenario would be continuously compounding interest, which compounds an infinite number of times. That sort of thing is possible; in fact, you'll learn to do it in precalculus.

The formula for calculating compound interest is slightly more complicated than simple interest; it looks like this:

$$b = p\left(1+\frac{r}{n}\right)^{nt}$$

In this formula, p is the principal investment, r is once again the annual interest rate in decimal form, n is the number of times interest is compounded in one year, and b is the balance in your account after exactly t years have elapsed.

Kelley's Cautions

Notice that the compound interest formula gives you the total *balance,* whereas the simple interest formula gives you the *interest* only—you had to add the principal to the interest in Example 1 in order to calculate the simple interest balance.

Example 2: How much more money would you make if you invested \$3,000 in a savings account whose 6.25% annual interest rate was compounded monthly rather than quarterly, if you planned on leaving the money alone for 18 months? (To keep our answers consistent, round all decimals to seven decimal places as you calculate.)

Solution: You'll have to calculate two separate balances, one with $n = 12$ for monthly compounding, and one with $n = 4$ for quarterly interest compounding. The other variables will match for both problems: $p = 3{,}000$, $r = .0625$, and $t = 1.5$. Be careful! The variable t must measure years, not months; since 18 months is exactly a year and a half, $t = 1.5$, not 18.

Kelley's Cautions

In both simple and compound interest problems, *t* must be measured in years. Therefore, if your investment collects interest for 24 months, $t = 2$, not 24, since 24 months equals two years.

Calculate the balance if you compound monthly.

$$\begin{aligned} b &= 3{,}000\left(1+\frac{.0625}{12}\right)^{(12)(1.5)} \\ &= 3{,}000(1+.0052083)^{18} \\ &= 3{,}000(1.0980173) \\ &= 3294.0519 \end{aligned}$$

Since banks don't award fractions of a penny, your final answer should only contain 2 decimal places: \$3,294.05. Now calculate the balance if interest is only compounded quarterly.

$$\begin{aligned} &= 3{,}000\left(1+\frac{.0625}{4}\right)^{4(1.5)} \\ &= 3{,}000(1+0.015625)^{6} \\ &= 3{,}000(1.0974893) \\ &= 3{,}292.4679 \end{aligned}$$

This time, your balance is \$3,292.47. Subtract the two balances to find the total difference: \$3,294.05 – \$3,292.47 = \$1.58. Sure, \$1.58 isn't a huge difference, but the larger the principal and the longer you leave the money in, the larger that difference will grow.

You've Got Problems

Problem 1: Calculate the *balance* of an account if its \$5,000 principal earns:

(a) Simple interest at an annual interest rate of 8.25% for 20 years.

(b) Interest compounded weekly ($n = 52$) at an annual interest rate of 8.25% for 20 years.

If necessary, round decimals to 7 places during your calculations.

Area and Volume Problems

Algebra teachers have a lot of weird tendencies. For one thing, they love fractals. I don't think I've ever met a math teacher who didn't have a poster featuring fractals proudly displayed in either his or her classroom, office, or wallet, next to pictures of his or her children. (In case you don't know, fractals are geometric shapes that look the same at any magnification. In other words, if you zoom in on the shape, its tiny detail looks just like the bigger shape. Practically speaking, a fractal usually looks like a psychedelic and groovy squiggle that should be featured on a tie-dyed T-shirt.)

Algebra teachers love geometric shapes so much, in fact, that they often ask you questions about them before you've taken a single geometry class. Don't be surprised if you see a few word problems regarding one of the following:

- **Area.** The amount of space covered by a two-dimensional object; the amount of carpet you'd need to cover a floor would depend on the floor's area.
- **Perimeter.** The distance around a two-dimensional object; the amount of fence you'd need to surround your yard equals the perimeter of your yard. (The perimeter of a circular object is called the *circumference*.)
- **Volume.** The amount of three-dimensional space inside an object; the amount of liquid inside a soda can represents the can's volume.
- **Surface area.** Measures the amount of "skin" needed to cover a three-dimensional object, neglecting its thickness; the amount of siding you'd need to completely cover a house represents the surface area of that house's walls.

There are lots of geometric formulas, most of which you'll learn in an actual geometry class. However, you may be expected to know the formulas listed in Table 19.1 by heart right now.

Table 19.1 Basic Geometric Formulas

Description	Formula	Variables
Area of a rectangle	$A = l \cdot w$	l = length, w = width
Perimeter of a rectangle	$P = 2(l + w)$	l = length, w = width
Area of a circle	$A = \pi r^2$	r = radius of circle
Circumference of a circle	$C = 2\pi r$	r = radius of circle
Volume of a rectangular solid	$V = l \cdot w \cdot h$	l = length, w = width, h = height
Volume of a cylinder	$V = \pi r^2 h$	r = radius; h = height
Surface area of a cube	$SA = 6l^2$	l = length of side

Basically, word problems in geometry come down to knowing the correct formula and plugging in values where they belong.

Example 3: (Inspired by *National Lampoon's Christmas Vacation*) Uncle Eddie is nervous about sledding with Clark W. Griswold, thanks to the rectangular metal plate in his head. (As it is, every time his wife uses the microwave, he forgets who he is for a half hour.) During the operation to implant the plate, right before the anesthetic kicked in, Eddie remembers hearing the doctor say that the plate's width is 3 centimeters shorter than its length. If the area of the plate is 54 cm^2, find its dimensions.

Critical Point

If a formula contains π, leave your answer in terms of π unless your instructor advises you otherwise. (In other words, your final answer should have π's in it.) For instance, the area of a circle with radius 5 will be $A = \pi \cdot 5^2 = 25\pi$.

Solution: You know nothing at all about the length of the rectangle, but you do know that the width equals the length minus 3 centimeters, so write that expression algebraically: $w = l - 3$. Because you know the area of the rectangle, you can then set up an equation based on the formula for rectangular area.

$$A = l \cdot w$$

Plug in $A = 54$ and $w = l - 3$.

$$54 = l(l-3)$$

When you distribute l, you end up with a quadratic equation that can be solved by factoring.

$$54 = l^2 - 3l$$

$$l^2 - 3l - 54 = 0$$

$$(l - 9)(l + 6) = 0$$

$$l = 9 \text{ or } -6$$

One of these answers doesn't make sense—the length of an object must *always* be positive, so throw out the –6 solution. You now know that the length of the rectangle must be 9 centimeters. Plug this value into the expression you came up with for width to determine the other missing dimension (also measured in centimeters).

$$w = l - 3$$

$$w = 9 - 3$$

$$w = 6$$

You've Got Problems

Problem 2: The height of a certain cylinder is exactly twice as large as its radius. If the volume of the cylinder is 36π in^3, what is the radius of the cylinder?

Speed and Distance Problems

Have you ever heard of a word problem like this one? "Train A heads north at an average speed of 95 miles per hour, leaving its station at the precise moment as another train, Train B, departs a different station, heading south at an average speed of 110 miles per hour. If these trains are inadvertently placed on the same track and start exactly 1,300 miles apart, how long until they collide?"

If that problem sounds familiar, it's probably because you watch a lot of television (like me). Whenever TV shows talk about math, it's usually in the context of a main character trying but failing miserably to solve the classic "impossible train problem." I have no idea why that is, but time and time again, this problem is singled out as the reason people hate math so much.

In fact, it's not so hard. This, like any distance and rate of travel problem, only requires one simple formula:

$$D = r \cdot t$$

CAUTION

Kelley's Cautions

Make sure that the units match in a travel problem. For instance, if the problem says you traveled at 70 miles per *hour* for 15 *minutes,* then $r = 70$ and $t = 0.25$. Since the speed is given in miles per *hour,* the time should be in hours also, and 15 minutes is equal to .25 hours. I got that decimal by dividing 15 minutes by the number of minutes in an hour: $\frac{15}{60} = \frac{1}{4} = 0.25$.

Distance traveled *(D)* is equal to your rate of speed *(r)* multiplied by the time *(t)* you traveled that speed. What makes most distance and rate problems tricky is that you usually have two things traveling at once, so you need to use the formula twice at the same time. In this problem, you'll use it once for Train A and once for Train B.

To keep things straight in your mind, you should use little descriptive subscripts. For example, use the formula $D_A = r_A \cdot t_A$ for Train A's distance, speed, and time values and use the formula $D_B = r_B \cdot t_B$ for Train B.

Example 4: Train A heads north at an average speed of 95 miles per hour, leaving its station at the precise moment as another train, Train B, departs a different station, heading south at an average speed of 110 miles per hour. If these trains are inadvertently placed on the same track and start exactly 1,300 miles apart, how long until they collide?

Critical Point

The little *A*'s in the formula $D_A = r_A \cdot t_A$ don't affect the values *D*, *r*, and *t*. They're just little labels to ensure that you only plug values corresponding to Train A into that formula.

Solution: Two trains means two distance formulas: $D_A = r_A \cdot t_A$ and $D_B = r_B \cdot t_B$. Your first goal is to plug in any values you can determine from the problem. Since Train A travels 95 mph, $r_A = 95$; similarly, $r_B = 110$.

Notice that the problem also says that the trains leave at the same time. This means that their travel times match exactly. Therefore, instead of denoting their travel times as t_A and t_B (which suggests they are different), I will write them both as t (which suggests they are equal). At this point, your formulas look like this:

$$D_A = 95t \qquad D_B = 110t$$

Here's the tricky step. The trains are heading toward one another on a track that's 1,300 miles long. Therefore, they must collide when, together, both trains have traveled a total of 1,300 miles. Of course, Train B is going to travel more of those 1,300 miles than Train A, since it's traveling faster, but that doesn't matter. You don't even have to figure out how far each train will go. All that matters is that when $D_A + D_B = 1300$, it's curtains. Luckily, you happen to know what D_A and D_B are ($95t$ and $110t$, respectively) so plug those into the equation and solve.

$$D_A + D_B = 1300$$
$$95t + 110t = 1300$$
$$205t = 1300$$
$$t \approx 6.341 \text{ hours}$$

So, the trains will collide in approximately 6.341 hours.

Kelley's Cautions

Even though you added the distances in this problem, you won't always do so—it depends on how the problem is worded. In Problem 3, for example, you will not calculate a sum.

You've Got Problems

Problem 3: Dave rode his bike from home to a 7-11 at an average speed of 17 mph, and the trip took 1.25 hours. However, as he pulled up to the store, he rode over some glass, causing both tires to go flat. Because of this rotten luck, he had to push his bike back home at an average speed of 3 mph. How long did the trip home take?

Mixture and Combination Problems

The final type of word problem I'll introduce involves mixing two or more things together into one, larger combination. What's actually being mixed doesn't matter, because the process is the same no matter what the ingredients.

Basically, your job will be to multiply the amount of each individual ingredient by some quantity that describes it, like its concentration, and then add those products, like so:

$$\boxed{\text{ingredient 1}} \cdot \boxed{\text{its value}} + \boxed{\text{ingredient 2}} \cdot \boxed{\text{its value}} = \boxed{\text{total weight}} \cdot \boxed{\text{its value}}$$

Notice that you set the sum equal to the total combined weight multiplied by its descriptive value.

Example 5: A bulk bag of trail mix, weighing 5 pounds, consists of 20% raisins by weight. The contents of the bag are mixed together with a smaller, 3-pound bag of trail mix that consists of 10% raisins by weight. What percentage of the resulting mixture is raisins?

Solution: If you weren't in algebra class, let's be honest, the correct answer to this question would be "Who cares? I don't even like raisins," but that answer doesn't get any partial credit, and it usually makes the teacher kind of angry.

In this problem the 5-pound bag will be ingredient 1, and its descriptive value is the percentage of raisins, expressed as a decimal: 20% = 0.20. Ingredient 2, then, will be the 3-pound bag, and its descriptive value will be 10% = 0.10. The total combined weight will be 3 + 5 = 8 pounds, but you don't know how many raisins are in the final mixture, so express that percentage with a variable, x. All together, your equation should look like this:

$$5(0.20) + 3(0.10) = 8(x)$$

CAUTION

Kelley's Cautions

If the descriptive value for an ingredient is a percentage, you should convert that to a decimal when creating your equation.

Multiply the values on the left side of the equation and add them together.

$$1 + 0.3 = 8x$$

$$1.3 = 8x$$

Solve for x by dividing both sides by 8.

$$\frac{1.3}{8} = x$$

$$x \approx 0.1625$$

To convert this decimal into a percentage, multiply it by 100; you'll get $x = 16.25\%$. Therefore, the giant, unappetizing 8-pound barrel of trail mix is 16.25% raisins. Hope you're hungry.

You've Got Problems

Problem 4: A 10-gallon fish tank currently contains 7 gallons of water with a 1.2% saline concentration. If you want the entire tank to have a saline concentration of 2%, what saline concentration should the final 3 gallons contain?

The Least You Need to Know

- Compound interest allows you to earn money based on the principal and interest earned to date, whereas simple interest is only earned on the principal.
- Geometric word problems require you to use algebra to calculate area, perimeter, volume, or surface area.
- Distance traveled is equal to the rate of travel times the time traveled.
- In mixture problems, you multiply the amount of each ingredient by some descriptive value, like its concentration.

Chapter 20

A Plethora of Practice

In This Chapter

- Measuring your understanding of all major algebra topics
- Practicing your skills
- Determining where you need more practice

Nothing helps you understand algebra like good, old-fashioned practice, and that's the purpose of this chapter. You can use it however you like, but I suggest one of the following three strategies:

1. As you finish reading each chapter, skip back here and work on the practice problems from that chapter.
2. If you're using this book as a refresher for an algebra class you've already taken, complete this test before you start reading the book. Then, go back and read the chapters containing problems you missed. Once you've reviewed those topics, try these problems again.
3. Save this chapter until the end, and use it to see how much you remember of each topic when you haven't seen it for a while.

Because these problems are just meant for practice, and not meant to teach new concepts, only the answers are given at the end of the chapter, without explanation or justification (unlike the problems in the "You've Got

Problems" sidebars throughout the book). However, these practice problems are designed to mirror those examples, so you can always go back and review if you forgot something or need extra review.

Are you ready? There's a lot of practice ahead of you—since some problems have multiple parts, there are actually over 110 practice problems in this chapter! (But no one said you have to do them all at one sitting.)

Chapter 1

1. Identify all the categories the number $0.6\overline{21}$, or .621212121…, belongs to.
2. Simplify $2 + (-3) - (-5) - (+7) + (+1)$.
3. Simplify.

   (a) $6 \times (-5)$

   (b) $-14 \div (-2)$
4. Simplify.

   (a) $4 + [3 - (8 + 2)]$

   (b) $|6 - 2(5 + 4)|$
5. Name the mathematical properties that guarantee each of the statements below is true.

   (a) $3 + (1 + 4) = (3 + 1) + 4$

   (b) $\frac{1}{2} \cdot 2 = 1$

   (c) $-8 + 0 = -8$

   (d) $2 \times 3 \times 5 = 3 \times 5 \times 2$

Chapter 2

6. Simplify the fraction $\frac{40}{64}$.
7. Rewrite the fractions so that they contain the least common denominator: $\frac{1}{2}, \frac{9}{4}, \frac{5}{6}$.

8. Simplify.

   (a) $\frac{2}{5}+\frac{7}{3}-\frac{1}{2}$

   (b) $\frac{5}{7}\cdot\frac{1}{10}\cdot\frac{2}{3}$

   (c) $\frac{6}{5}\div\frac{1}{3}$

Chapter 3

9. Translate into mathematical expressions:

   (a) Seventeen less than a number

   (b) The product of a number and one more than three times that number

10. Evaluate the expressions.

    (a) $(-2)^3$

    (b) 5^4

11. Simplify the expression $\left(\frac{x^3y^{-2}}{x^{-1}}\right)^4$.

12. Write the numbers in scientific notation.

    (a) .00000679

    (b) 23,400,000,000,000,000,000,000,000

13. Apply the distributive property: $-3x(4x^5 + 7x - 9)$.

14. Simplify the expression: $10 - (12 - 4 \cdot 2)^2$.

15. Evaluate the expression $3x(xy - y^2)$ if $x = 2$ and $y = -3$.

Chapter 4

16. Solve the equations.

    (a) $3 + w = 12$

    (b) $-\frac{7}{5}x = \frac{2}{15}$

    (c) $7y - 19 = 2y - 4$

    (d) $4(x - 5) + 8 = -28$

    (e) $2|x-3|+5=13$

17. Solve the equation $5x - 4y = 20$ for y.

Chapter 5

18. Identify the points indicated on the coordinate plane below.

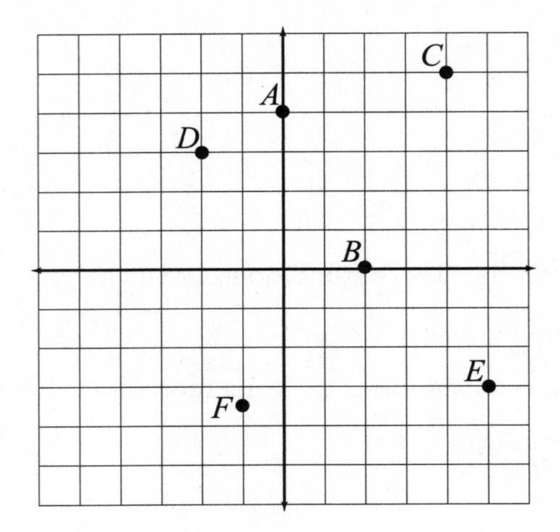

19. What are the coordinates of the intercepts for the linear equation $4x - 3y = 8$?

20. Which of the equations below has the following graph?

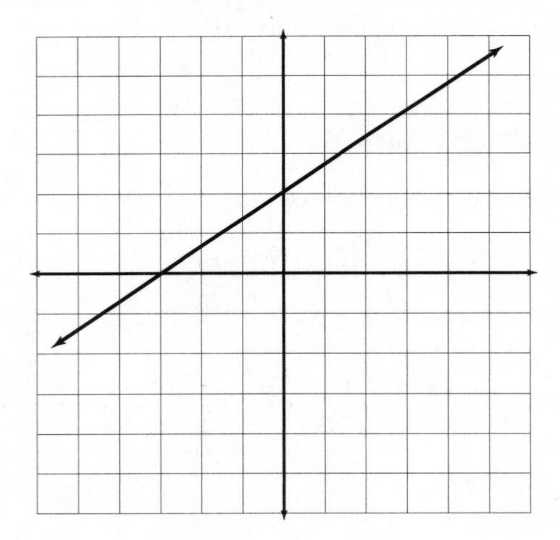

(A) $2x + 3y = 6$

(B) $-2x + 3y = 6$

(C) $2x - 3y = 6$

(D) $2x + 3y = -6$

21. Calculate the slope of the line that passes through the points (4,3) and (–9,2).

22. What coordinate pair represents the vertex of the graph of the equation $y = -|x - 4| - 5$?

Chapter 6

23. Determine the equation of the line with slope $m = -\frac{2}{3}$ that passes through the point (–1,2) and write it in standard form.

24. Determine the slope and the y-intercept of the line with equation $5x - 3y = -9$.

25. Write the equation of the line that passes through the points (–3,7) and (4,1) in standard form.

26. Write the equation of line m in standard form if m passes through the point (7,0) and is perpendicular to the line $2x + y = -6$.

Chapter 7

27. Solve the inequality $14 - 3(y + 5) > 8$.

28. Which of the following is the correct graph of the inequality $-3 \le 2x + 5 \le 1$?

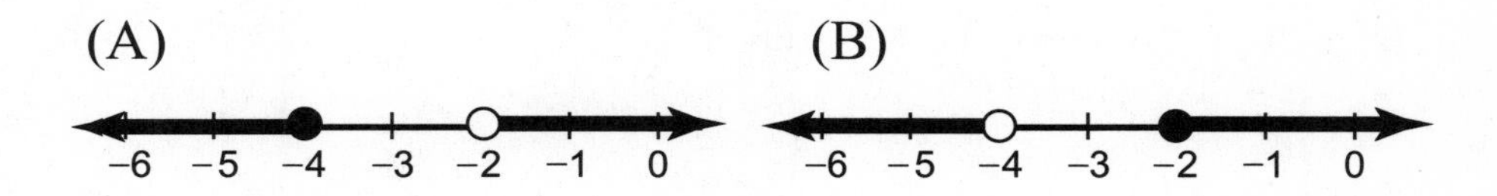

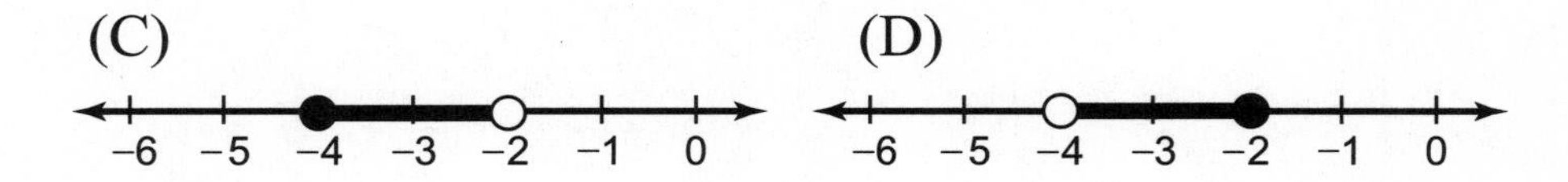

29. Solve the inequalities.

(a) $|2x - 1| > 0$

(b) $|x + 4| \le 2$

30. Write the equation of the inequality in standard form whose graph is pictured below.

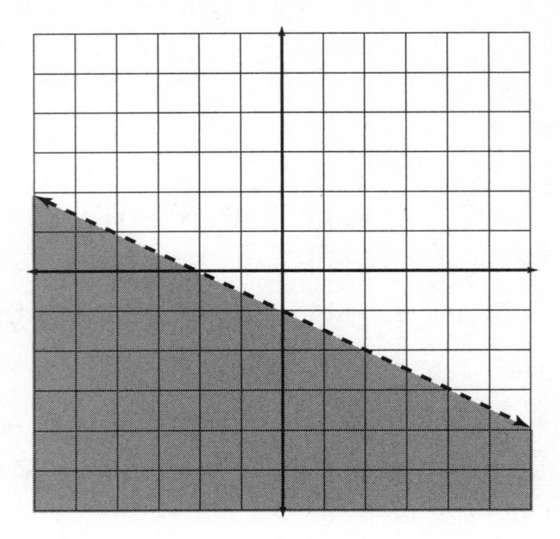

Chapter 8

31. Solve the system by graphing the equations.

$$\begin{cases} 2x - y = 1 \\ x + 3y = 18 \end{cases}$$

32. Solve the system using substitution.

$$\begin{cases} 3x + y = 1 \\ 4x + 3y = -2 \end{cases}$$

33. Solve the system using elimination.

$$\begin{cases} 4x + 3y = -1 \\ -2x + 15y = 6 \end{cases}$$

34. Describe the solution for the system of equations.

$$\begin{cases} x - 3y = -4 \\ 3x - 9y = -11 \end{cases}$$

35. Graph the solution for the system of inequalities.

$$\begin{cases} y \leq 4x - 3 \\ y > -\frac{1}{3}x + 1 \end{cases}$$

Chapter 9

36. If $A = \begin{bmatrix} 4 & -1 \\ 3 & 0 \\ -2 & 9 \end{bmatrix}$ and $B = \begin{bmatrix} -3 & 1 \\ 6 & -2 \\ 12 & 5 \end{bmatrix}$, calculate $2A - B$.

37. Calculate $\begin{bmatrix} 7 & -3 \\ -2 & 1 \end{bmatrix} \cdot \begin{bmatrix} -1 & 0 & 4 & -6 \\ -8 & 5 & 1 & 2 \end{bmatrix}$.

38. Evaluate the determinants.

 (a) $\begin{bmatrix} -6 & -5 \\ -1 & 9 \end{bmatrix}$

 (b) $\begin{bmatrix} 2 & -3 & -1 \\ 1 & -4 & 0 \\ 5 & 1 & 6 \end{bmatrix}$

39. Use Cramer's Rule to solve the system of equations.

$$\begin{cases} 9x + 2y = 7 \\ 36x - 3y = -16 \end{cases}$$

Chapter 10

40. Classify the polynomials.

 (a) $-7x^4 - 9$

 (b) $19x^3$

41. Simplify the expression: $3x^2(2xy + 8x - 3y) - 2xy(y^3 + 2x^2 - 7x)$.

42. Calculate the products and simplify.

 (a) $(2x^2y - 3xy^3)(xy + 7y^3)$

 (b) $(x - 3y)(x^2 + 2xy - y^2)$

43. Calculate the quotients.

 (a) $(x^4 + 3x^3 - x + 5) \div (x^2 - 2x + 1)$

 (b) $(x^4 - 9x^3 + 25x^2 - 26x + 17) \div (x - 5)$

Chapter 11

44. Factor the polynomials.

    (a) $12x^3y - 6x^2y^2 + 9x^5y^3$

    (b) $12x^2 + 18x - 10xy - 15y$

    (c) $16x^2 - 1$

    (d) $x^3 - 8$

    (e) $x^2 + 9x - 36$

    (f) $6x^2 - 17x - 3$

Chapter 12

45. Simplify the radicals.

    (a) $\sqrt[3]{-27x^{10}y^8}$

    (b) $\sqrt{75x^2y^3}$

46. Rewrite the expression $8^{7/3}$ as an integer.

47. Simplify the radical expressions, rationalizing if necessary.

    (a) $\sqrt{12x} - \sqrt{48x}$

    (b) $\left(\sqrt[3]{9x^2y}\right)\left(\sqrt[3]{6x^2y^2}\right)$

    (c) $\sqrt{35x^2y^7} \div \sqrt{7xy^8}$

48. Solve the equation: $\sqrt[3]{x+7} = 2$.

49. Simplify the expressions.

    (a) i^{11}

    (b) $\sqrt{-18x}$

50. Simplify the expressions.

    (a) $(3 - 2i) - 5(-2 + 3i)$

    (b) $(-3 + i)(5 - 6i)$

    (c) $(2 + i) \div (4 - i)$

Chapter 13

51. Solve the equation by factoring: $5(x^2 - 2x) = -13x^2 - x + 14$.

52. Solve by completing the square: $3x^2 + 12x - 21 = 0$.

53. Solve using the quadratic formula: $2x^2 + 5x + 6 = 0$.

54. Without calculating them, determine how many real solutions the equation $9x^2 - 3x = -\frac{1}{4}$ has.

55. Solve the inequality and graph the solution: $x^2 + 2x - 24 \leq 0$.

Chapter 14

56. Demonstrate that $x - 1$ is a factor of the polynomial $2x^3 - 11x^2 + 4x + 5$ and use that information to completely factor the trinomial.

57. Given that 8 is a root of the equation $x^3 - 5x^2 - 42x + 144 = 0$, use that information to find the other two roots.

58. Identify all four rational roots of the polynomial equation $4x^4 - 13x^3 - 37x^2 + 106x - 24 = 0$.

59. Find all the roots of the equation: $x^3 + 7x^2 + 19x + 45 = 0$.

Chapter 15

60. Calculate the following if $f(x) = x^3 + 4x - 1$ and $g(x) = x^2 + 6$.

    (a) $(f - g)(x)$

    (b) $(fg)(-2)$

61. Calculate the following if $h(x) = (x+1)^2$ and $j(x) = x - 3$.

    (a) $h(j(x))$

    (b) $(j \circ h)(-1)$

62. If $f(x) = 6(x - 1) + 2$, find $f^{-1}(x)$.

63. Given $g(x)$ as defined below, evaluate: (a) $g(-2)$ and (b) $g(3)$.

$$g(x) = \begin{cases} |2x+3|, & x \leq -2 \\ 2x+3, & -2 < x < 2 \\ -|2x+3|, & x \geq 2 \end{cases}$$

Chapter 16

64. Sketch the function $f(x)=\sqrt{-x}+2$ using transformations, and then verify the graph's accuracy by regraphing the function, this time plotting points.

65. Examine the graph of $p(x)$ below. Based upon its graph alone, can $p(x)$ be classified as a function? If so, is $p(x)$ one-to-one?

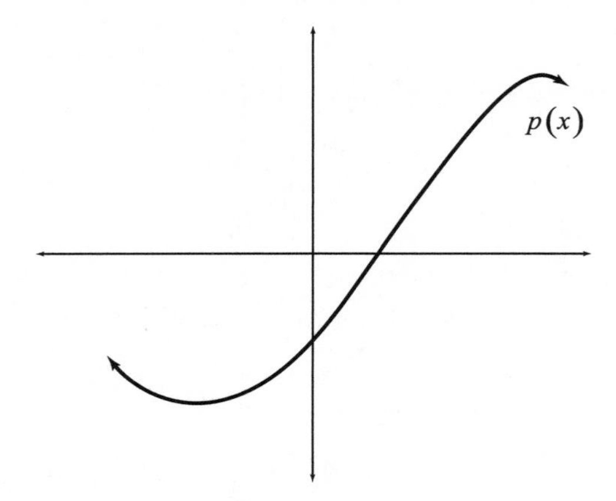

66. Given the graph of $k(x)$ below, determine the function's domain and range.

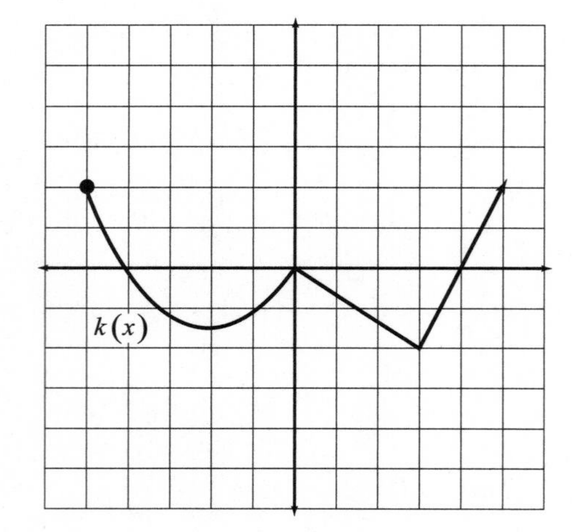

Chapter 17

67. Simplify the expression $\dfrac{3x^2+13x-10}{2x^2+11x+5}$.

68. Rewrite each as a single expression and simplify.

(a) $\dfrac{3x}{x^2-9x+20}-\dfrac{2}{x-5}$

(b) $\dfrac{x^2y^3}{x^2+13x+36}\cdot\dfrac{x^2-81}{x^4y^2}$

(c) $\dfrac{x^2-6x-27}{x^2-11x+18}\div\dfrac{x^3+2x^2-3x}{x^2-2x}$

69. Simplify the complex fraction.

$$\frac{\dfrac{6x+4}{x^2}}{\dfrac{12x+8}{x^3-x}}$$

Chapter 18

70. Solve the equations.

(a) $\dfrac{x}{2x^2-5x+3}-\dfrac{x+5}{2x^2+x-6}=\dfrac{6}{x^2+x-2}$

(b) $\dfrac{x+6}{18}=\dfrac{-3}{x-9}$

71. The variables x and y vary proportionally, and when $x = 3$, $y = 5$. Determine the value of x when $y = 16$.

72. Assume x varies inversely with y, and that $y = 35$ when $x = 5$. Find the value of y when $x = 25$.

73. Solve and graph the inequality: $\dfrac{3x+1}{x-4}\le 2$.

Chapter 19

74. How much simple interest will a principal investment of $9,500 earn in a savings account with a fixed annual interest rate of 4.95% if it is left alone for 35 years?

75. Exactly 20 years ago, you invested a big sum of money in a savings account with a 6.8% annual interest rate compounded quarterly. If your current balance is $55,852.71, what was your principal? (Hint: Solve for p in the compound interest equation.)

76. You want to mark the perimeter of your rectangular yard with fencing, and you bought exactly the correct amount of fencing to accomplish the task: 362 feet. If the length of your yard is exactly 1 foot longer than twice its width, find the yard's *length*.

77. The hare wants a rematch. Learning from his past mistakes pitted against the tortoise, he will hold off on the victory laps until the race is over. If the tortoise gets a head start of 90 minutes and travels at an average velocity of 3 feet/minute, how long will it take the hare (who travels at an average velocity of 600 feet/minute) to overtake the tortoise?

78. Sixteen ounces of an apple-flavored drink containing 15% real apple juice is mixed with 64 ounces of another juice drink. If the resulting mixture is 12.6% apple juice, what concentration of juice did the 64-ounce drink contain?

Solutions

Chapter 1: (1) Positive, real, rational, complex (see Chapter 12); (2) –2; (3a) –30; (3b) 7; (4a) –3; (4b) 12; (5a) The associative property for addition; (5b) Multiplicative inverse property; (5c) Identity property for addition; (5d) Commutative property for multiplication.

Chapter 2: (6) $\frac{5}{8}$; (7) $\frac{6}{12}, \frac{27}{12}, \frac{10}{12}$; (8a) $\frac{67}{30}$; (8b) $\frac{1}{21}$; (8c) $\frac{18}{5}$.

Chapter 3: (9a) $x - 17$; (9b) $x(3x + 1)$; (10a) –8; (10b) 625; (11) $\frac{x^{16}}{y^8}$; (12a) 6.79×10^{-6}; (12b) 2.34×10^{25}; (13) $-12x^6 - 21x^2 + 27x$; (14) –6; (15) –90.

Chapter 4: (16a) $w = 9$; (16b) $x = -\frac{2}{21}$; (16c) $y = 3$; (16d) $x = -4$; (16e) $x = -1$ or 7; (17) $y = \frac{5}{4}x - 5$.

Chapter 5: (18) $A = (0,4)$, $B = (2,0)$, $C = (4,5)$, $D = (-2,3)$, $E = (5,-3)$, and $F = \left(-1, -3\frac{1}{2}\right)$ or $F = \left(-1, -\frac{7}{2}\right)$; (19) (2,0) and $\left(0, -\frac{8}{3}\right)$; (20) B; (21) $\frac{1}{13}$; (22) (4,–5).

Chapter 6: (23) $2x + 3y = 4$; (24) slope $= \frac{5}{3}$, y-intercept $= 3$; (25) $6x + 7y = 31$; (26) $x - 2y = 7$.

Chapter 7: (27) $y < -3$; (28) D; (29a) $x > \frac{1}{2}$ or $x < \frac{1}{2}$ (in other words, the solution is all real numbers except $x = \frac{1}{2}$); (29b) $-6 \le x \le -2$; (30) $x + 2y < -2$.

Chapter 8: (31) (3,5); (32) (1,–2); (33) $\left(-\frac{1}{2}, \frac{1}{3}\right)$; (34) Inconsistent (no solution); (35) See graph that follows.

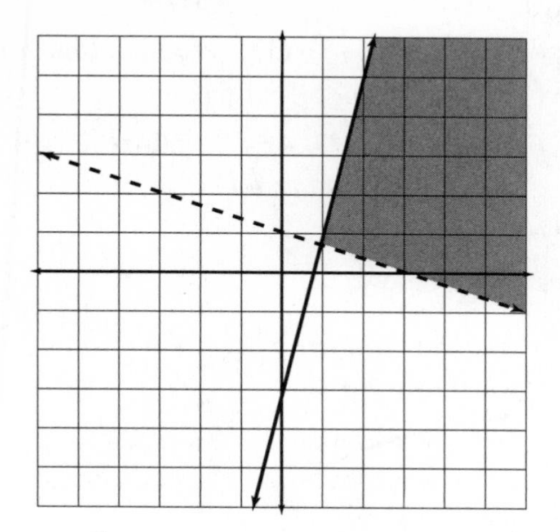

Chapter 9: (36) $\begin{bmatrix} 11 & -3 \\ 0 & 2 \\ -16 & 13 \end{bmatrix}$; (37) $\begin{bmatrix} 17 & -15 & 25 & -48 \\ -6 & 5 & -4 & 14 \end{bmatrix}$; (38a) –59; (38b) –51; (39) $\left(-\frac{1}{9}, 4\right)$.

Chapter 10: (40a) Quartic binomial; (40b) Cubic monomial; (41) $2x^3y + 24x^3 + 5x^2y - 2xy^4$; (42a) $2x^3y^2 + 11x^2y^4 - 21xy^6$; (42b) $x^3 - x^2y - 7xy^2 + 3y^3$; (43a) $x^2 + 5x + 9 + \frac{12x - 4}{x^2 - 2x + 1}$; (43b) $x^3 - 4x^2 + 5x - 1 + \frac{12}{x - 5}$.

Chapter 11: (44a) $3x^2y(4x - 2y + 3x^2y^2)$; (44b) $(6x - 5y)(2x + 3)$; (44c) $(4x + 1)(4x - 1)$; (44d) $(x - 2)(x^2 + 2x + 4)$; (44e) $(x + 12)(x - 3)$; (44f) $(6x + 1)(x - 3)$.

Chapter 12: (45a) $-3x^3y^2\left(\sqrt[3]{xy^2}\right)$; (45b) $5|xy|\sqrt{3y}$; (46) 128; (47a) $-2\sqrt{3x}$; (47b) $3xy\sqrt[3]{2x}$; (47c) $\frac{\sqrt{5xy}}{y}$; (48) $x = 1$; (49a) $-i$ (49b) $3i\sqrt{2x}$; (50a) $13 - 17i$; (50b) $-9 + 23i$; (50c) $\frac{7}{17} + \frac{6i}{17}$.

Chapter 13: (51) $x = -\frac{2}{3}, x = \frac{7}{6}$; (52) $x = -2 + \sqrt{11},\ x = -2 - \sqrt{11}$; (53) $x = -\frac{5}{4} + \frac{i\sqrt{23}}{4}$, $x = -\frac{5}{4} - \frac{i\sqrt{23}}{4}$; (54) One; (55) $-6 \le x \le 4$, see graph below.

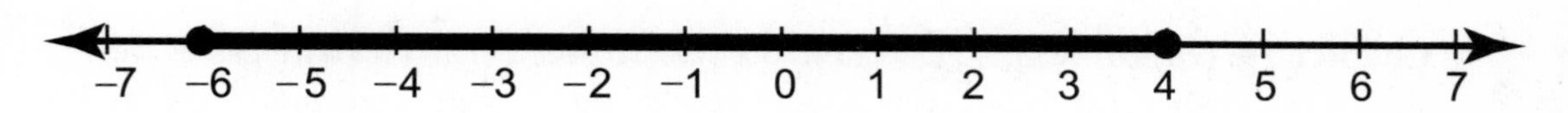

Chapter 14: (56) $(x - 1)(2x + 1)(x - 5)$; (57) –6 and 3; (58) $-3, \frac{1}{4}, 2, 4$; (59) $-5, -1+2i\sqrt{2}, -1-2i\sqrt{2}$.

Chapter 15: (60a) $(f - g)(x) = x^3 - x^2 + 4x - 7$; (60b) $(fg)(-2) = -170$; (61a) $h(j(x)) = x^2 - 4x + 4$; (61b) $(j \circ h)(-1) = -3$; (62) $f^{-1}(x) = \frac{1}{6}x + \frac{2}{3}$; (63a) $g(-2) = 1$; (63b) $g(3) = -9$.

Chapter 16: (64) See graph below.

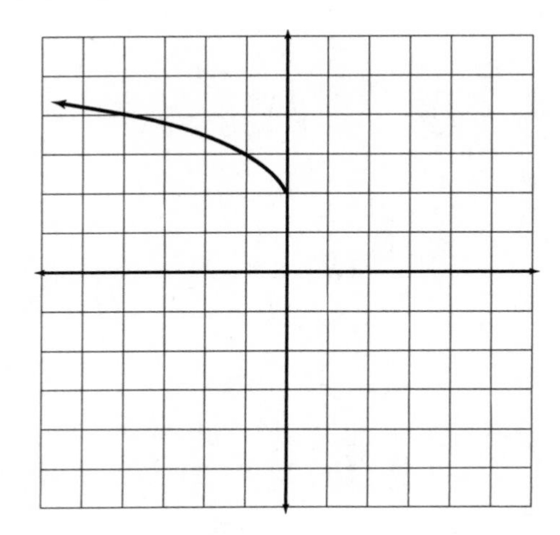

(65) $p(x)$ is a function, but is not one-to-one; (66) Domain: $x \geq -5$, range: $y \geq -2$.

Chapter 17: (67) $\frac{3x-2}{2x+1}$; (68a) $\frac{x+8}{(x-5)(x-4)}$; (68b) $\frac{y(x-9)}{x^2(x+4)}$; (68c) $\frac{1}{x-1}$; (69) $\frac{x^2-1}{2x}$.

Chapter 18: (70a) $x = \frac{23}{14}$; (70b) $x = 0,3$; (71) $x = \frac{48}{5}$; (72) $y = 7$; (73) $-9 \leq x < 4$, see graph below.

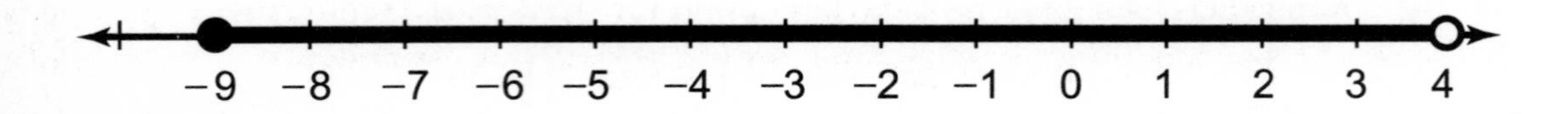

Chapter 19: (74) \$16,458.75; (75) \$14,500; (76) 121 feet; (77) The hare passes the tortoise after just $\frac{9}{20}$ of a minute—if you like, set up the proportion $\frac{9}{20} = \frac{x}{60}$ and solve for x to find out that $x = 27$ seconds; (78) 12%.

Solutions to "You've Got Problems"

Here are detailed solutions for all the "You've Got Problems" sidebars throughout the book. I suggest you only turn back here to look at the answers once you've tried your best and either got an answer or are hopelessly stuck; you should find just enough information to get you through any important or tricky steps.

Don't just read the problem and then flip back here to read the answer! Unless you actually *do the problem yourself first*, you'll never master the concept on your own.

Chapter 1

1. Positive, rational, and real numbers only. Because $\frac{3}{7}$ is a fraction, it is automatically excluded from the groups of natural numbers, whole numbers, and integers. Although it is positive, it cannot be considered even, odd, prime, or composite, because those four classifications only apply to integers.
2. 7. Eliminate double signs to get $6 + 2 - 5 + 4$. You earn a total of 12 $(6 + 2 + 4)$ but lose 5 for a final answer of 7.

3. (a) 40. 8 times 5 equals 40, and since the numbers are both negative (and thus have the same sign), the answer is positive.

   (b) –5. The answer must be negative since the signs are different.

4. $-(8) = -8$; $|8| = 8$.

5. (a) 10. Subtract (4 – 2) first to get 5 × 2.

   (b) 2. The innermost symbols, the parentheses, should be done first to get $|2-4|$. Subtract correctly to get $|-2|$, and then take the absolute value.

6. (a) Commutative property of addition (the order of the numbers changes).

   (b) Additive inverse property (the result is 0, the identity element for addition).

   (c) Associative property of multiplication (the order of the numbers doesn't change, just the grouping).

Chapter 2

1. (a) $\frac{1}{3}$; divide 7 and 21 by the GCF, which is 7.

   (b) $\frac{3}{5}$; GCF = 8.

2. $\frac{10}{30}$, $\frac{25}{30}$, and $\frac{21}{30}$. The least common denominator is 30. Multiply the numerator and denominator of the first fraction by 10, the second fraction by 5, and the third by 3.

3. $\frac{5}{4}$. Start with common denominators $\left(\frac{18}{12}-\frac{4}{12}+\frac{1}{12}\right)$, combine the numerators $\left(\frac{15}{12}\right)$, and then simplify by dividing the numerator and denominator by 3.

4. 1. First write 8 as a fraction $\left(\frac{5}{6}\times\frac{8}{1}\times\frac{3}{20}\right)$, then multiply numerators together and denominators together to get $\frac{120}{120}$. Any number divides into itself exactly 1 time.

5. $\frac{3}{2}$. Remember, do the inside of the parentheses first, using a common denominator of 14: $\frac{7}{14}-\frac{4}{14}=\frac{3}{14}$. That leaves the division problem $\frac{3}{14}\div\frac{1}{7}$, which is equivalent to the multiplication problem $\frac{3}{14}\times\frac{7}{1}=\frac{21}{14}$. Simplify.

Chapter 3

1. $\frac{1}{3}x-5$. One third of a number equals $\frac{1}{3}x$, and 5 less than that is $\frac{1}{3}x-5$.

2. 81. $(-3)(-3)(-3)(-3) = 9(-3)(-3) = -27(-3) = 81$.

3. x^9y^{11}. According to Rules 3 and 4 you'll get $(x^{15}y^5) \cdot (x^{-6}y^6)$. Then, apply Rule 1 $(x^{15+(-6)}y^{5+6})$ and simplify.

4. (a) 2.3451×10^{13}.

   (b) 1.25×10^{-9}

5. $12xy^3 - 30y^5 + 48y^3$. Remember, if you're multiplying exponential expressions with the same base, you can add the powers. Otherwise, just list the variables next to one another in alphabetical order.

6. 12. You've got exponents, division, and multiplication, so start with exponents: $100 \div 25 \cdot 3$. Since multiplication and division are in the same step, work from left to right: $4 \cdot 3 = 12$.

7. 80. Substitute in the values to get $5((-1)-3)^2$. Combine $-1 - 3$ to get -4 inside the parentheses and the resulting expression is $5(-4)^2$. Since $(-4)^2 = (-4)(-4) = 16$, the answer will be $5 \cdot 16$.

Chapter 4

1. $x = 11$. To isolate x on the left side, add -8 to both sides of the equation (since -8 is the opposite of 8); that leaves the expression $19 - 8$ on the right side, which should be simplified.

2. $w = -20$. Multiply both sides of the equation by $-\frac{5}{4}$ to get $\frac{20}{20}w = -\frac{80}{4}$, which simplifies to $w = -20$.

3. (a) $x = \frac{17}{6}$. Distribute the 3 to get $6x - 3 = 14$, and add 3 to both sides to separate the variable: $6x = 17$. Divide both sides by 6 (the fraction cannot be simplified).

   (b) $x = -10$. On both sides of the equation, subtract $4x$ and add 7 to get $-2x = 20$. Divide both sides by -2 and simplify.

4. $x = -5$ and $x = 15$. Isolate the absolute value quantity by adding 6 to both sides, then split into two equations ($x - 5 = 10$ and $x - 5 = -10$) and solve them separately.

5. $y = -3x + \frac{5}{3}$. Isolate the y term by subtracting $9x$ on both sides: $3y = -9x + 5$.

   Now divide both sides by 3: $y = \frac{-9x+5}{3}$. Write as separate fractions $\left(y = \frac{-9}{3}x + \frac{5}{3}\right)$ and simplify.

Chapter 5

1. $A = (-4,-4)$, $B = (0,3)$, $C = (-2,0)$, $D = (-5,4)$, $E = (2,-1)$, and $F = \left(5\frac{1}{2}, 1\frac{1}{2}\right)$ or $F = \left(\frac{11}{2}, \frac{3}{2}\right)$.

2. Solve for y to get $y = 4x + 2$. Plugging in $x = -1$, 0, and 1 results in the coordinate pairs $(-1,-2)$, $(0,2)$, and $(1,6)$. Plot and connect those points.

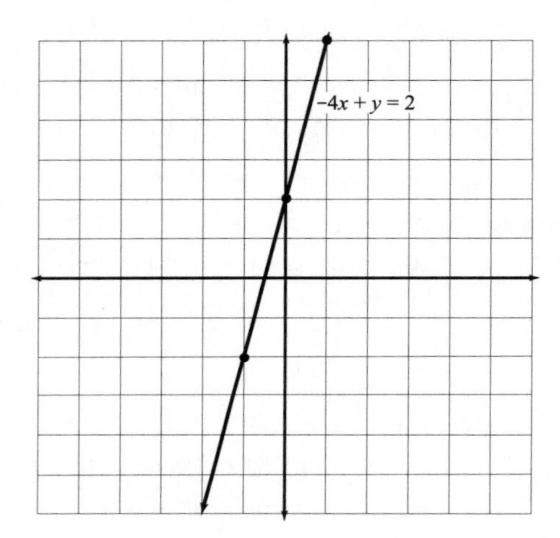

3. Plug in 0 for x to get $2y = -8$, and a y-intercept of $(0,-4)$; plug in 0 for y to get $4x = -8$, and an x-intercept of $(-2,0)$. Plot the points and connect them to get the graph.

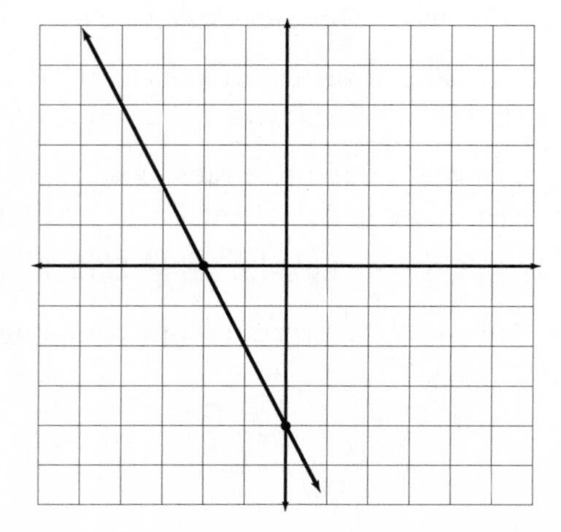

4. $-\frac{2}{3}$. Apply the slope formula: $\frac{6-(0)}{-5-(4)} = \frac{6}{-9}$ and simplify the answer.
5. Set $2x - 4 = 0$ and solve to get $x = 2$. Plug this into the equation to get the corresponding y to complete the vertex's ordered pair: (2,1). Now choose an x less than 2 and an x greater than 2 (perhaps $x = 0$ and $x = 4$) and calculate the coordinate pairs associated with them. For the x-values I indicated, those pairs are (0,5) and (4,5). Draw a line beginning at the vertex and passing through each of those two points.

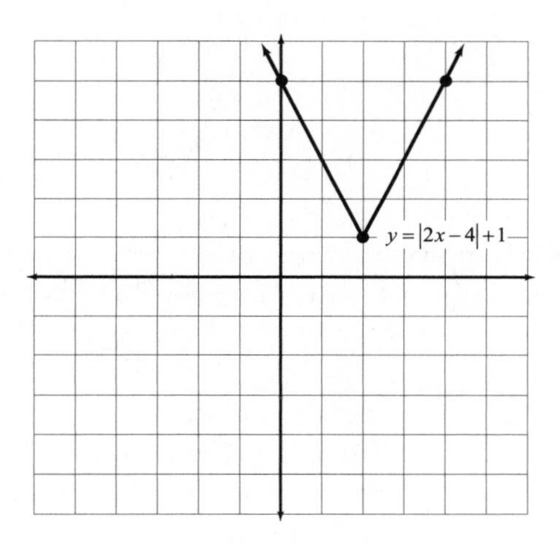

Chapter 6

1. $y = 4x - 15$. Substitute $m = 4$, $x_1 = 2$, and $y_1 = -7$ into the point-slope formula to get $y - (-7) = 4(x - 2)$. Simplify both sides ($y + 7 = 4x - 8$) and finish by subtracting 7 from both sides.

2. Slope $= -\frac{3}{2}$; y–intercept = (0,2). Subtract $3x$ from both sides and divide by 2 to solve for y: $y = -\frac{3}{2}x + 2$. The slope is the coefficient of the x term, and the y-intercept is the constant.

3. Solve for y to get $y = -2x - 1$. Note that slope *must* be a fraction, but since -2 is an integer, it is also rational: $-\frac{2}{1}$. Either count down two units and right one from the y-intercept of (0,–1), or count up two units and left one to get another point on the line.

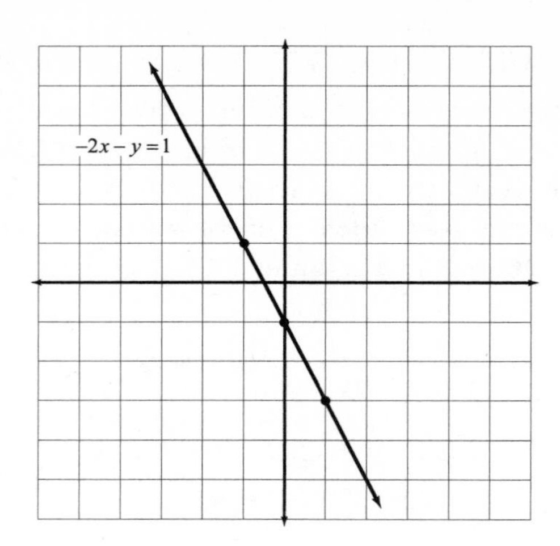

4. $2x - 15y = -28$. Multiply everything by the least common denominator of 12 to eliminate fractions: $15y = 28 + 2x$. Move the $2x$ term by subtracting it from both sides: $-2x + 15y = 28$. Finally, multiply everything by –1 to make the x-term positive.
5. $x - y = -8$. First calculate the slope: $m = \frac{0-(5)}{-8-(-3)} = \frac{-5}{-5} = 1$. Now apply the point-slope formula using one of the points. If you choose the point (–8,0), you get $y = 1(x - (-8))$, or $y = x + 8$. Put this equation in standard form.
6. $x - 3y = -9$. Put $-2x + 6y = 7$ in slope intercept form $\left(y = \frac{1}{3}x + \frac{7}{6}\right)$ to determine its slope. Line j must have the same slope ($\frac{1}{3}$), so apply the point-slope form using the given point (–6,1) to get $y - 1 = \frac{1}{3}(x - (-6))$. Put that equation in standard form, as indicated by the problem.

Chapter 7

1. (a) False; no number can be less than itself; had the statement been $-3 \le -3$, it would have been true, though, since the values on either side of the inequality are equal.

   (b) True; 5 is either less than *or* equal to 11 (only one of those two conditions can be true at a time, and it's true that 5 is less than 11).

2. $w \le 15$. First, simplify the left side by distributing the 2: $2w - 12 \le 18$. Isolate the variable by adding 12 to both sides: $2w \le 30$. Divide both sides by 2 to get your final answer. No need to reverse the inequality sign, since you're dividing both sides by a positive number.

3. Solve the inequality by subtracting $2x$ and 1 from both sides to get $-3x \geq -6$. Now divide both sides by -3 to get $x \leq 2$. The graph will be a solid dot at 2 and an arrow extending from there to the left.

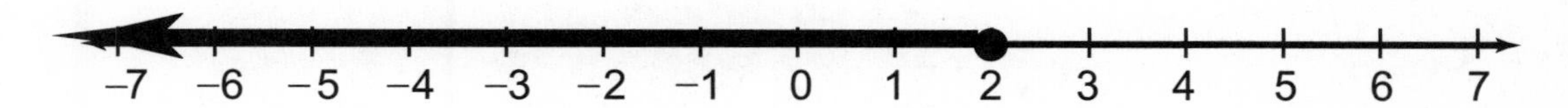

4. $-3 < x < 4$. Subtract 5 from all three parts to get $-6 < 2x < 8$, and then divide everything by 2 to get the final answer.

5. Place a solid dot at -2 and an open dot at 5 on the number line, and connect the dots with a dark line.

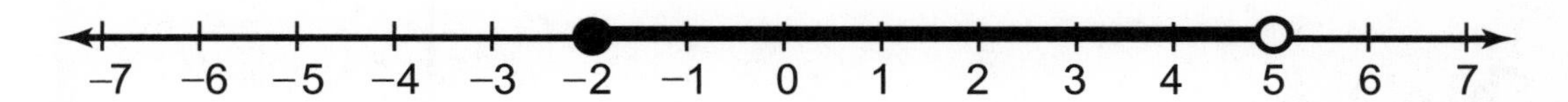

6. $3 < x < 7$. Divide both sides by 4 to isolate the absolute value quantity: $|x - 5| < 2$. Rewrite the statement as $-2 < x - 5 < 2$ and solve by adding 5 to each part of the inequality. The graph is a line segment with open endpoints at 3 and 7.

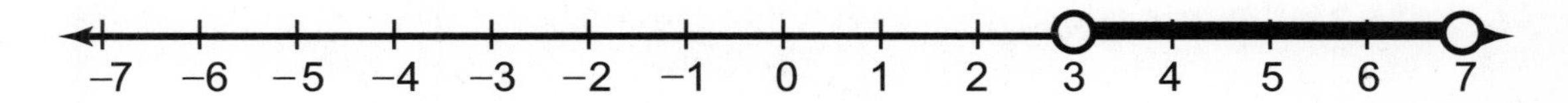

7. $x \geq 6$ or $x \leq 2$. Split the inequality appropriately ($x - 4 \geq 2$ or $x - 4 \leq -2$) and solve. Since both inequality symbols specify "or equal to," make sure to use solid dots on the graph.

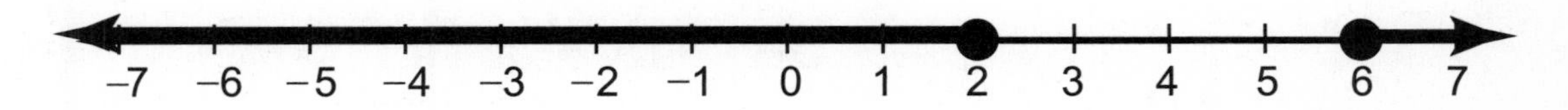

8. It's easy to graph the line using the slope-intercept technique of Chapter 6. The inequality sign specifies "or equal to," so the line should be solid. If you test the point (0,0), you get $0 \geq 3$, which is *false*, so you should shade the region *not* containing the origin.

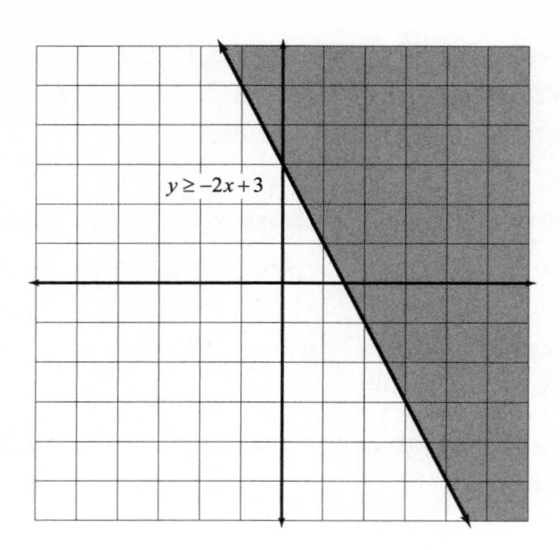

Chapter 8

1. (0,3). Both graphs intersect at that point on the y-axis.

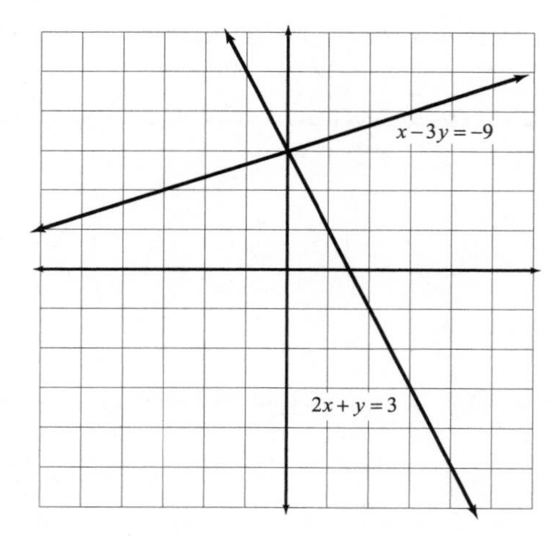

2. (3,–2). Notice that the first equation has an x-coefficient of 1, so solve it for x to get $x = 4y + 11$. Plug that in for x in the other equation: $3(4y + 11) + 7y = -5$. Solve the equation, and you get $y = -2$. Plug this y-value into the equation you solved for x to get $x = -4(2) + 11$, and solve that equation to get the matching x-value of 3.

3. (–2,7). Multiply the first equation by 2 to get $4x - 2y = -22$; when you add this to the second equation, you'll get $9x = -18$. Solve that and plug the result, $x = -2$, into one of the original equations to get the final answer.

You could also have multiplied the first equation by 5 (resulting in $10x - 5y = -55$) and multiplied the second equation by –2 (resulting in $-10x - 4y = -8$). When you add those equations together, you get $-9y = -63$ (or $y = 7$). Find the x-value by plugging $y = 7$ into one of the equations.

4. Dependent. Multiply the first equation by 2 to eliminate either x or y, and add the equations together. Every term will cancel out, resulting in the true statement $0 = 0$, which indicates a dependent system.

5. Both equations are graphed easily using the slope-intercept form. You'll end up shading to the left of the dotted graph of $y > 4x - 5$ and below the solid graph of $y \le -\frac{2}{3}x + 1$ to get the graph below.

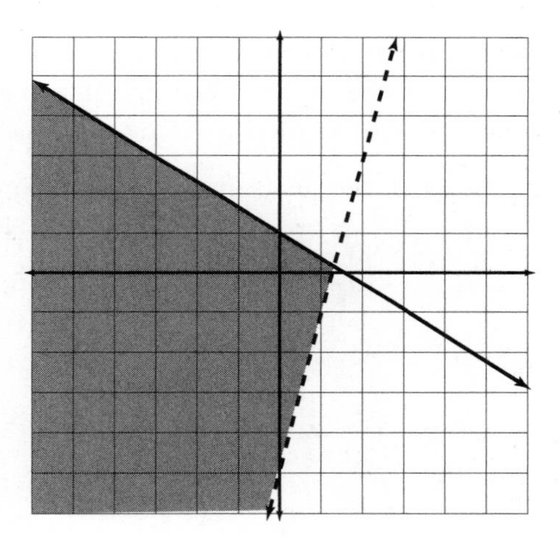

Chapter 9

1. $4B = \begin{bmatrix} 4 & 16 & 36 \\ -8 & 28 & -12 \\ 44 & -20 & 24 \end{bmatrix}$. Just multiply each element of B by 4 to get $4B$.

2. $4A - 3B = \begin{bmatrix} -7 & 14 \\ 41 & -6 \end{bmatrix}$. Multiply A by a scalar of 4 and B by a scalar of –3 to get $4A - 3B = \begin{bmatrix} -4 & 2 \\ 20 & 12 \end{bmatrix} + \begin{bmatrix} -3 & 12 \\ 21 & -18 \end{bmatrix}$. Combine corresponding matrix elements to get the final answer.

3. The product $B \cdot A$ exists because the number of columns in B equals the number of rows in A. The order of $B \cdot A$ is 3×4; the same number of rows as B, and the same number of columns as A.

4. $\begin{bmatrix} 17 & -19 \\ -19 & 9 \\ -24 & 8 \end{bmatrix}$. Moving your fingers deftly, you'll get the following products and sums; simplify correctly to get the final answer.

$$\begin{bmatrix} 2(6)+(-5)(-1) & 2(-2)+(-5)(3) \\ (-3)(6)+1(-1) & (-3)(-2)+1(3) \\ -4(6)+0(-1) & -4(-2)+0(3) \end{bmatrix}$$

5. 3. Multiply 9 times –1 and then subtract the product of 3 and –4: $9(-1)-3(-4)$. According to the order of operations, multiply first to get to $-9 + 12$, and then combine the numbers to get the final answer.

6. 279. Multiplying along the diagonals gives you this expression:

$$(-2)(-5)(-3) + (-4)(1)(7) + (9)(3)(2) - (7)(-5)(9) - (2)(1)(-2) - (-3)(3)(-4) = -30 - 28 + 54 + 315 + 4 - 36$$

7. $\left(-\frac{1}{3}, 1\right)$. Your matrices will be $A = \begin{bmatrix} -7 & -5 \\ 1 & 2 \end{bmatrix}$, $B = \begin{bmatrix} 6 & -7 \\ 3 & 1 \end{bmatrix}$, and $C = \begin{bmatrix} 6 & -5 \\ 3 & 2 \end{bmatrix}$.

The determinants will be $|A| = -9$, $|B| = 27$, and $|C| = 27$. Therefore, $x = \frac{-9}{27}$ and $y = \frac{27}{27}$; simplify the fractions.

Chapter 10

1. (a) Because of its two terms and degree of 3, this is a cubic binomial.

   (b) If x is not raised to an explicit power, then the power is understood to be 1. So, the polynomial can be rewritten as $11x^1$. Since it has one term with a variable raised to a power of 1, it's a linear monomial.

2. $5x^2 - 12x + 1$. Start by distributing the 2 to get $3x^2 - 9 + 2x^2 - 12x + 10$. In this polynomial, $3x^2$ and $2x^2$ are like terms, as are –9 and 10, so combine them. The $-12x$ term has no like terms in the polynomial, so leave it alone. It's best to write your answer in decreasing exponential form (standard form).

3. $15x^5y + 12x^4y^2 - 6x^2y^6$. Distribute $3x^2y$ to each term and calculate the product; $(3x^2y)(5x^3) = 15x^5y$, $(3x^2y)(4x^2y) = 12x^4y^2$, and $(3x^2y)(-2y^5) = -6x^2y^6$.

4. $2x^2 - 5xy - 3y^2$. Distribute the $2x$ and the y each through the expression $(x - 3y)$, and you'll get

$$(2x)(x) + (2x)(-3y) + (y)(x) + (y)(-3y) = 2x^2 - 6xy + xy - 3y^2$$

Notice that $-6xy$ and xy are like terms, so simplify accordingly: $-6xy + xy = -5xy$.

5. $x - 11 + \frac{52}{x+4}$. Here are the steps:

$$\begin{array}{r|l} & \quad x - 11 \\ \hline x+4 & x^2 - 7x + 8 \\ & \underline{-x^2 - 4x} \\ & \quad -11x + 8 \\ & \quad \underline{11x + 44} \\ & \qquad\quad 52 \end{array}$$

6. $4x^2 + 6x + 2 + \frac{5}{x-2}$. Your synthetic division should look like this:

$$\begin{array}{r|rrrr} 2 & 4 & -2 & -10 & 1 \\ & & 8 & 12 & 4 \\ \hline & 4 & 6 & 2 & 5 \end{array}$$

Chapter 11

1. $3x^3y^2(3x^2 + xy - 2y^5)$. The GCF of the coefficients is 3. The highest power of x contained in every term is 3, and the highest power of y in every term is 2. So, you should divide every term in the original polynomial by $3x^3y^2$; the results are contained within the parentheses in the answer.

   Note that when you subtract equal exponential powers $\left(\frac{y^2}{y^2} = y^{2-2} = y^0 = 1\right)$, it's the same as multiplying by 1, so that variable cancels out and disappears.

2. $(2x + 1)(6x^3 + 7)$. The first two terms have a GCF of $6x^3$ and the second have a GCF of 7. Factor them out to get this: $6x^3(2x + 1) + 7(2x + 1)$. (Since the GCF of 7 gives you matching binomials of $2x + 1$, you don't have to worry about factoring out a -7 instead.) To finish, factor out the common binomial and write what's left in parentheses.

3. $5(x + 5)(x - 5)$. Start by factoring out the GCF of 5 to get $5(x^2 - 25)$. Since $x^2 - 25$ is the difference of perfect squares (if $a = x$ and $b = 5$), it must be factored further.

4. $2x(x - 8)(x - 4)$. Sorry if the hint ruined the surprise for you; since the GCF is $2x$, factor it out to get $2x(x^2 - 12x + 32)$. Since 32 is positive, the signs of the mystery numbers match, and since x's coefficient is negative, both numbers must be negative also. The only two negative numbers that add up to –12 and multiply to 32 are –8 and –4.

5. $(4x - 1)(x + 6)$. No GCF exists. The two numbers that add up to 23 and have a product of –24 are: 24 and –1. Rewrite the polynomial as $4x^2 + (24 - 1)x - 6$ and distribute the x to get $4x^2 + 24x - x - 6$; factor by grouping to finish.

Chapter 12

1. $10x^3y\sqrt{3y}$. Since the radical's a square root, only perfect squares escape. Rewrite the coefficient as $100 \cdot 3$, or $10^2 \cdot 3$. Notice that x^6 is a perfect square since it's equivalent to $x^3 \cdot x^3$. Rewrite it as $(x^3)^2$, so it's raised to the second power. The best you can do for y^3 is to rewrite it as $y^2 \cdot y$, so it contains one perfect square: $\sqrt{100^2 \cdot 3 \cdot \left(x^3\right)^2 \cdot y^2 \cdot y}$.
2. 125. Rewrite $25^{3/2}$ as $\left(\sqrt{25}\right)^3$. Since $\sqrt{25} = \sqrt{5^2} = 5$, then $\left(\sqrt{25}\right)^3 = 5^3 = 125$.
3. $6x\sqrt[3]{x}$. Rewrite the first radical as $\sqrt[3]{2^3 \cdot x^3 \cdot x}$, so that it contains the most possible perfect cubes. Once those perfect cubes are paroled, the radical simplifies to $2x\sqrt[3]{x}$, and the original problem becomes $2x\sqrt[3]{x} + 4x\sqrt[3]{x}$. Now you have like radicals, whose coefficients are like terms, so add those coefficients together and follow the result with the like radical.

4. $6|xy|\sqrt{x}$. Multiply the radicands to get $\sqrt{36x^3y^2}$ and simplify the radical. Don't forget both x and y need to be in absolute value signs outside the radical.

5. $\dfrac{\sqrt{xy}}{3|x|}$. Simplify the fraction $\sqrt{\dfrac{2x^2y^3}{18x^3y^2}}$ to get $\sqrt{\dfrac{y}{9x}} = \dfrac{\sqrt{y}}{3\sqrt{x}}$. Rationalize by multiplying both the numerator and denominator by $\sqrt{x}$ to get $\dfrac{\sqrt{xy}}{3\sqrt{x^2}}$ and simplify.

6. $x = 19$. Divide both sides by 2 to isolate the radical: $\sqrt{x-3} = 4$. Since the radical has index 2, raise both sides to the second power: $\left(\sqrt{x-3}\right)^2 = 4^2$, which gives you $x - 3 = 16$. Solve the equation by adding 3 to both sides.

7. $5i$. Pull the negative sign out of the square root as i and the perfect square 36 as well to get $6i + i^3$. Rewrite i^3 so that it includes a power of 2: $6i + i^2 \cdot i$. Finally, replace i^2 with –1 to get $6i + (-1)i = 6i - i$. Treat $6i$ and $-i$ as like terms and subtract their coefficients.

8. (a) $c + d = 3 + 8 - 4i + i = 11 - 3i$.

   (b) $c - d = 3 - 4i - (8 + i) = 3 - 8 - 4i - i = -5 - 5i$.

   (c) $c \cdot d = 24 + 3i - 32i - 4i^2 = 24 - 29i - 4(-1) = 24 - 29i + 4 = 28 - 29i$.

   (d) $\frac{3-4i}{8+i} \cdot \frac{8-i}{8-i} = \frac{24-35i+4i^2}{64-i^2} = \frac{20-35i}{65} = \frac{20}{65} - \frac{35}{65}i = \frac{4}{13} - \frac{7}{13}i$.

Chapter 13

1. $x = -\frac{5}{2}, 0, \frac{5}{2}$. Subtract $25x$ from both sides and factor out the GCF to get $x(4x^2 - 25) = 0$. The quantity in parentheses is a difference of perfect squares, so factor it: $x(2x + 5)(2x - 5) = 0$. Now, set each factor equal to 0, including the GCF, x: $x = 0$ or $2x + 5 = 0$ or $2x - 5 = 0$ and solve the final two equations. (The first equation, $x = 0$, is already solved.)

2. $x = -3 \pm 2\sqrt{3}$. The coefficient of x^2 is already 1, so add 3 to both sides to get $x^2 + 6x = 3$. According to the bug, you should add the square of half of 6, or $3^2 = 9$ to both sides: $x^2 + 6x + 9 = 3 + 9$, which is equivalent to $(x + 3)^2 = 12$. Take the square root of both sides to get $x + 3 = \pm 2\sqrt{3}$ and solve for x.

3. $x = -3 \pm 2\sqrt{3}$. In this problem, $a = 1$, $b = 6$, and $c = -3$.

$$x = \frac{-6 \pm \sqrt{36 - 4(1)(-3)}}{2 \cdot 1}$$

$$x = \frac{-6 \pm \sqrt{48}}{2}$$

$$x = \frac{-6 \pm 4\sqrt{3}}{2}$$

$$x = \frac{-6}{2} \pm \frac{4\sqrt{3}}{2}$$

$$x = -3 \pm 2\sqrt{3}$$

4. 1. Evaluate the discriminant: $b^2 - 4ac$ will equal 1600 – 4(25)(16), which simplifies to 0. Therefore the quadratic equation has only one real solution, and it's a double root.

5. $x \leq -3$ or $x \geq \frac{1}{2}$. Factor the quadratic to get $(2x - 1)(x + 3)$, which yields the critical numbers $x = \frac{1}{2}$ and $x = -3$. They split the number line into the intervals $x \leq -3$, $-3 \leq x \leq \frac{1}{2}$, and $x \geq \frac{1}{2}$. Choose one test value from within each interval

(for instance –4, 0, and 1) and test in the inequality. Valid solutions are found in both the intervals $x \leq -3$ and $x \geq \frac{1}{2}$; because a correct solution can come from either interval, use the word "or" when writing the answer.

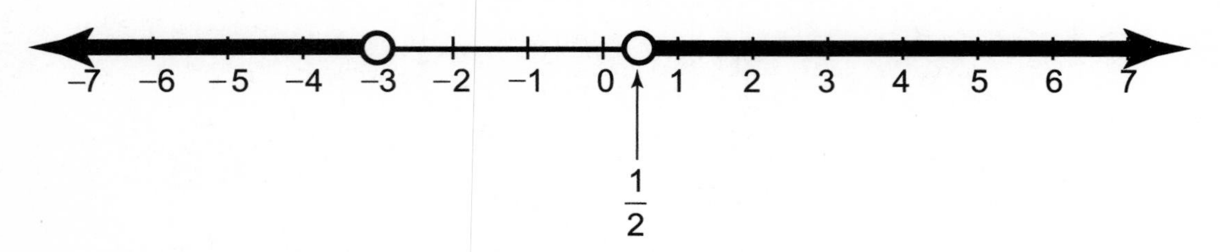

Chapter 14

1. $(x + 2)(x - 2)(2x + 5)$. You can use synthetic division, since $x + 2$ is linear.

2	2	5	–8	–20
		–4	–2	20
	2	1	–10	0

   Therefore, $2x^3 + 5x^2 - 8x - 20 = (x + 2)(2x^2 + x - 10)$. The quadratic can be further factored into $(2x + 5)(x - 2)$.

2. –9 and 1. If –2 is a root, then $(x - (-2))$, or $(x + 2)$, is a factor of the polynomial.

2	1	10	7	–18
		–2	–16	18
	1	8	–9	0

   Fully factor the equation to get $(x + 2)(x + 9)(x - 1) = 0$ and set each factor equal to 0.

3. Roots: $-2, -\frac{1}{2}, \frac{1}{2}$, and 1. When fully factored, the polynomial looks like $(x - 1)(x + 2)(2x - 1)(2x + 1) = 0$.

4. $3, -\frac{1}{4} + \frac{i\sqrt{55}}{4}$, and $-\frac{1}{4} - \frac{i\sqrt{55}}{4}$. The only rational root is 3; synthetic division gives you a factored form of $(x - 3)(2x^2 + x + 7) = 0$. Use the quadratic formula to solve the equation formed when the quadratic factor is set equal to 0: $x = \dfrac{-1 \pm \sqrt{-55}}{4}$.

Chapter 15

1. Start by evaluating the functions at $x = -1$.

$(h+k)(x) = x^2 + 11x - 8$ $(h+k)(-1) = -18$	$(h-k)(x) = x^2 + 3x - 2$ $(h-k)(-1) = -4$
$(hk)(x) = 4x^3 + 25x^2 - 41x + 15$ $(hk)(-1) = 77$	$\left(\frac{h}{k}\right)(x) = \frac{x^2 + 7x - 5}{4x - 3}$ $\left(\frac{h}{k}\right)(x) = \frac{-11}{-7} = \frac{11}{7}$

The order, from least to greatest, is $(h + k)(-1)$, $(h - k)(-1)$, $\left(\frac{h}{k}\right)(-1)$, $(hk)(-1)$. Note that these functions are *much* easier to evaluate if you determine that $h(-1) = -11$ and $k(-1) = -7$, and then just add, subtract, multiply, and divide those numbers.

2. (a) $f\left(\sqrt{x-2}\right) = x + 3$. Plug $\sqrt{x-2}$ in for x in the expression $x^2 + 5$ to get $f\left(\sqrt{x-2}\right) = \left(\sqrt{x-2}\right)^2 + 5$. The radical and exponent will cancel, resulting in $f\left(\sqrt{x-2}\right) = x - 2 + 5$. Combine like terms.

   (b) $g\left(x^2+5\right) = \sqrt{x^2+3}$. Plug $x^2 + 5$ in for x in the expression $\sqrt{x-2}$ to get $g\left(x^2+5\right) = \sqrt{\left(x^2+5\right) - 2}$. Combine like terms. Note that the expression $\sqrt{x^2+3}$ cannot be further simplified. While it's true that x^2 is a perfect square, you can only release it from the radical prison if it's *multiplied* by the other contents of the radicand, not added.

3. When composed with one another, the functions should cancel out, leaving just x.

$$\boxed{f(g(x))}$$

$$= \frac{3}{1}\left(\frac{x+5}{3}\right) - 5$$

$$= \frac{3x+15}{3} - 5$$

$$= \frac{3}{3}x + \frac{15}{3} - 5$$

$$= x + 5 - 5$$

$$= x$$

$$\boxed{g(f(x))}$$

$$= \frac{(3x-5)+5}{3}$$

$$= \frac{3x}{3}$$

$$= x$$

4. $g^{-1}(x) = \frac{1}{7}x + \frac{3}{7}$. Rewrite the $g(x)$ with a y and reverse that y with the x to get $x = 7y - 3$. Solving for y gives you $y = \frac{x+3}{7}$, or $y = \frac{1}{7}x + \frac{3}{7}$. Replace y with $g^{-1}(x)$ to finish.

5. Least to greatest: $f(3), f(1), f(-1)$. The function values will be

$$f(-1) = 3(-1) + 4 = 1,\ f(1) = 1 - (1)^2 = 0,\text{ and } f(3) = 3 - (3)^2 = -6.$$

Chapter 16

1. As you learned in Chapter 5, the graph has a V shape.

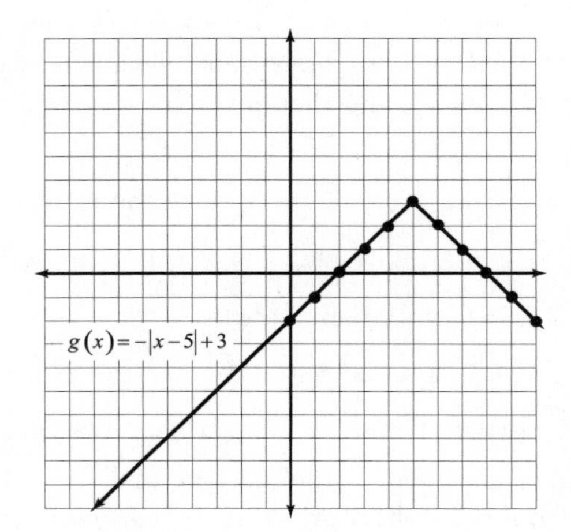

2. According to the vertical line test, $g(x)$ is a function; however, g fails the horizontal line test (any horizontal line $y = c$ will pass through the graph twice if $c < 3$), so it isn't one-to-one. Unfortunately, g is allergic to apples, so the answer to the final question is no.

3. The graph extends infinitely to the right and left, and any vertical line will intersect it. Therefore, the domain of $g(x)$ is all real numbers (nothing is disqualified). However, only horizontal lines drawn at a height of 3 or below will hit the graph, so the range is $y \le 3$.

4. Since you're plugging in $-x$ instead of just x, the graph of x^3 is reflected about the y-axis. Additionally, move the graph down two units.

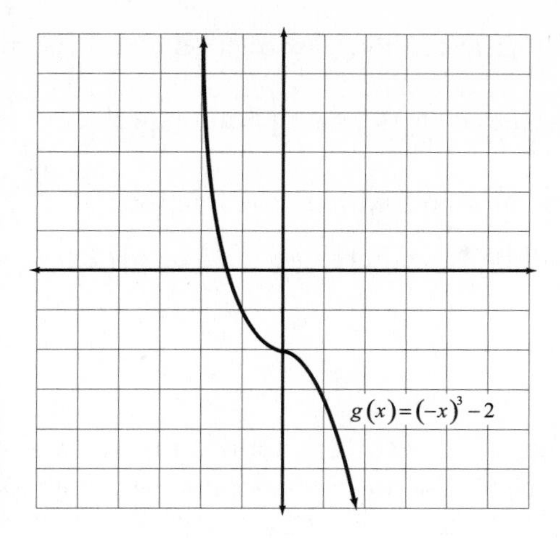

Chapter 17

1. $\frac{x^2+3x}{3x+5}$. Factor both the numerator and denominator completely to get $\frac{x(x+3)(2x-1)}{(2x-1)(3x+5)}$. Eliminate the common factor of $(2x-1)$ to get $\frac{x(x+3)}{(3x+5)}$. You can leave your answer like that or multiply the terms in the numerator.

2. $\frac{x^2+3x+16}{(x-2)(x+2)(x+5)}$. Factor the denominators to get $\frac{x}{(x+2)(x-2)}+\frac{8}{(x-5)(x-2)}$.

   The LCD will be $(x+2)(x-2)(x-5)$. Multiply the top and bottom of each fraction by the terms needed to reach the least common denominator.

   $$\frac{(x-5)}{(x-5)}\cdot\frac{x}{(x+2)(x-2)}+\frac{8}{(x-5)(x-2)}\cdot\frac{(x+2)}{(x+2)}=\frac{x^2-5x+8x+16}{(x-2)(x+2)(x+5)}$$

3. $\frac{(x+2)}{(x-4)(x+3)}$. Take the reciprocal of the second fraction, factor everything, and multiply the fractions together:

   $$\frac{(x-6)(x+2)(3x+2)}{(3x+2)(x-4)(x+3)(x-6)}$$

   Simplify the fraction.

4. $(x - 4)(x + 2)$. Rewrite the complex fraction as a division problem, and then use the reciprocal to rewrite it as a multiplication problem: $\frac{(x^2 - 7x + 12)(x^2 + 4x + 4)}{(x+2)(x-3)}$.
Factor the numerator and simplify the fraction: $\frac{\cancel{(x-3)}(x-4)\cancel{(x+2)}(x+2)}{\cancel{(x+2)}\cancel{(x-3)}}$. The denominator will be 1, so there's no need to write it in your answer.

Chapter 18

1. No solution. Factor the second denominator into $(x - 8)(x + 2)$ and then multiply the entire equation by the least common denominator, which is $(x - 8)(x + 2)$. Once you simplify, you'll get $(x + 2)(x + 3) + x = (x - 8)(x + 2)$. Multiply the binomial pairs: $x^2 + 5x + 6 + x = x^2 - 6x - 16$. When you set the entire equation equal to 0, you end up with $11x + 22 = 0$; solve for x, and you get $x = -2$. However, if you plug that into the original equation, both fractions will have 0 in the denominator, so it is not a valid solution. Therefore, there are no valid solutions at all.

2. $x = -2$ or 1. Cross multiply to get $(3x - 2)(x + 1) = x \cdot 2x$, which gives you $3x^2 + x - 2 = 2x^2$. Subtract $2x^2$ from both sides to get $x^2 + x - 2 = 0$, which can be solved by factoring: $(x + 2)(x - 1) = 0$.

3. $y = \frac{175}{4}$ (or 43.75). Find k, the constant of proportionality: $k = \frac{y}{x} = \frac{15}{12} = \frac{5}{4}$. Now use the equation $\frac{y}{x} = k$ again, this time plugging in the k you found and $x = 35$: $\frac{y}{35} = \frac{5}{4}$. Cross multiply and solve: $4y = 175$.

4. 337.5. You know that $xy = k$, so find k: $(9)(75) = k = 675$. Now solve the equation $xy = k$ for x when $y = 2$ and $k = 675$.

$$x \cdot 2 = 675$$
$$x = \frac{675}{2} = 337.5$$

5. $x < 2$ or $x > \frac{13}{2}$. Subtract 3 from both sides, rewrite it as $\frac{3}{1}$, and multiply its numerator and denominator by $x - 4$, the least common denominator: $\frac{x+7}{x-2} - \frac{3}{1} \cdot \frac{x-2}{x-2} < 0$. Combine the fractions and simplify to get $\frac{x+7-3(x-2)}{x-2} < 0$, which simplifies to $\frac{-2x+13}{x-2} < 0$. The critical numbers (mark both on the number line with open dots) are $x = 2$ and $x = \frac{13}{2}$ (or 6.5). Test intervals to get a final solution of $x < 2$ or $x > \frac{13}{2}$.

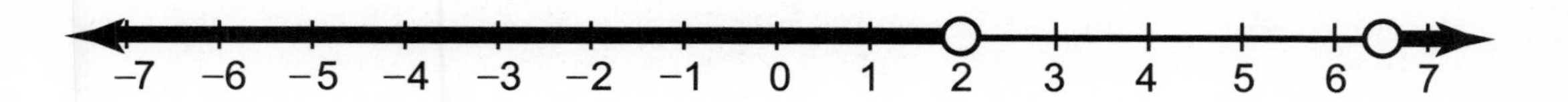

Chapter 19

1. (a) \$13,250. Plug $p = 5{,}000$, $r = .0825$, and $t = 20$ into the simple interest formula to get $i = (5{,}000)(.0825)(20) = 8{,}250$. Add the interest to the principal to get the balance: \$8,250 + \$5,000 = \$13,250.

   (b) \$25,999.84. Use the compound interest formula with $p = 5{,}000$, $r = .0825$, $t = 20$, and $n = 52$ to get $b = 5{,}000\left(1 + \frac{.0825}{52}\right)^{(52)(20)}$ and simplify.

2. $r = \sqrt[3]{18}$ inches. Since the height is twice the radius, you can write $h = 2r$. You're given the volume of the cylinder, so use the corresponding formula, $V = \pi r^2 h$, and plug in the known values: $36\pi = \pi r^2(2r)$. Simplify the right side to get $36\pi = 2\pi r^3$. Solve the equation for r by first dividing both sides by 2π: $18 = r^3$. (Notice that, like any other number, π cancels out when divided by itself.) Finally, take the cube root of both sides to finish.

3. 7.083 hours (that's quite a hike). Think of the trip to the store as trip A and the trip home as B. Use two distance formulas, $D_A = r_A \cdot t_A$ and $D_B = r_B \cdot t_B$. Set $r_A = 17$ and $t_A = 1.25$, so $D_A = 17(1.25) = 21.25$, meaning that the distance between home and the 7-11 is 21.25 miles! Now, for the trip home. His speed is 3 mph, so $r_B = 3$. The distance home will be the exact same value as the distance to the store, so $D_B = 21.25$. Therefore, the formula $D_B = r_B \cdot t_B$ becomes $21.25 = 3 \cdot t_B$. Solve for t_B to get your final answer.

4. The 3 gallons should have a salinity concentration of approximately 3.87%. Ingredient 1 is the 7 gallons of water with a descriptive value of 1.2% = 0.012. Ingredient 2 is the 3 gallons of water that will fill the tank, but you don't know its saline value, so use the variable x to represent it. The total volume of the tank is 10 gallons; that number should be multiplied by 2% = 0.02, creating the equation $7(0.012) + 3x = 10(0.02)$. Solve for x.

$$\begin{aligned} 0.084 + 3x &= 0.2 \\ 3x &= 0.2 - 0.084 \\ 3x &= 0.116 \\ x &\approx 0.0387 \end{aligned}$$

   Convert 0.0387 to a percent: 3.87%.

Appendix B

Glossary

abscissa Fancy-pants word for the x part of a coordinate pair.

absolute value The positive value of the indicated number or expression.

area The amount of space covered by a two-dimensional object.

asymptote Boundary line that a graph gets infinitely close to but never actually touches.

axiom *See* property.

base (of an exponential expression) In the expression x^2, the base is x; x will be multiplied by itself two times.

bomb method Technique used to factor quadratic polynomials whose leading coefficient are not equal to 1; also called factoring by decomposition.

coefficient The number appearing at the beginning of a monomial; the coefficient of $12xy^2$ is 12.

common denominator A denominator shared by one or more fractions; it must be present in order to add or subtract those fractions.

completing the square Process used to solve quadratic equations by forcing them to contain binomial perfect squares.

complex fraction A fraction that contains (as its numerator, denominator, or both) another fraction; also called a compound fraction.

complex number Has form $a + bi$, where a and b are real numbers; any imaginary or real numbers are both automatically complex as well.

composite A number which has factors in addition to itself and 1.

composition of functions The process of inputting one function into another; it can be denoted with a circle operator: $(f \circ g)(x) = f(g(x))$.

compound fraction *See* complex fraction.

compound inequality One inequality statement that is actually the combination of two others, such as $a < x < b$.

compound interest Method of earning interest on an entire balance, rather than just the initial investment.

conjugate The quantity $a - bi$ associated with every complex number $a + bi$; the only difference between a complex number and its conjugate is the sign immediately preceding their imaginary part, bi.

constant A number with no variable attached to it.

constant of proportionality Real number that describes either direct or indirect variation.

coordinate pair The point (x,y) used to describe a location in the coordinate plane.

coordinate plane Grid used to visualize mathematical graphs.

Cramer's Rule Technique used to solve systems of equations using determinants of matrices.

critical numbers The values of x for which an expression equals 0 or is undefined.

cross multiplication Method of solving proportions in which you multiply the numerator of one fraction by the denominator of the other and set those products equal.

cube root A radical whose index is 3.

degree The largest exponent in a polynomial.

denominator The bottom number in a fraction.

dependent Describes a system of equations with an infinite number of solutions.

determinant (of a matrix) Real number value defined for square matrices only.

direct variation Exhibited when two values, x and y, have the property $y = k \cdot x$, where k is a real number called the constant of proportionality.

discriminant The expression $b^2 - 4ac$; it describes, based on its sign, how many real number solutions the equation has.

dividend The quantity *a* in the division problem $b\overline{)a}$.

divisible If *a* is divisible by *b*, then $\frac{a}{b}$ is an integer; in other words, there is no remainder.

divisor The quantity *b* in the division problem $b\overline{)a}$.

domain The set of possible inputs for a function.

double root Solution repeated once in a polynomial equation; it's the result of a repeated factor in the polynomial.

elements The numbers within a matrix; in advanced matrix problems, elements may include variables, expressions, or even other matrices. Also called entries.

equation A mathematical sentence including an equal sign.

even A number that is divisible by 2.

exponent In the expression x^2, the exponent is 2; *x* will be multiplied by itself two times.

expression Mathematical incomplete sentence that doesn't contain an equal sign.

factor If *a* is a factor of *b*, then *b* is divisible by *a*.

fraction Ratio of two numbers representing some portion of an integer.

Fundamental Theorem of Algebra Guarantees that a polynomial of degree *n*, if set equal to 0, will have exactly *n* roots.

function A relation whose inputs each have a single, corresponding output.

greatest common factor The largest factor of two or more numbers or terms.

grouping symbols Elements like parentheses and brackets that explicitly tell you what to simplify first in a problem.

horizontal line test Tests the graph of a function to determine whether or not it's one-to-one.

i The imaginary value $\sqrt{-1}$.

identity element The number (0 for addition, 1 for multiplication) that leaves a number's value unchanged when the corresponding operation is applied.

imaginary number Has form *bi*, where *b* is a real number and $i = \sqrt{-1}$.

improper fraction A fraction whose numerator is greater than its denominator.

inconsistent Describes a system of equations that has no solution.

index In the radical expression $\sqrt[a]{b}$, a is the index.

indirect variation Exhibited by two quantities, x and y, when their product remains constant even as the values of x and y change: $xy = k$; also called inverse variation.

inequality A statement whose two sides are either definitely not equal (if the symbol is $<$ or $>$) or possibly unequal (if the symbol is $\leq$ or $\geq$).

integer A number with no obvious fraction or decimal part.

intercept Point on the x- or y-axis through which a graph passes.

interval Segment of the number line, as defined by an inequality's critical numbers.

inverse functions Functions which cancel each other out when composed with one another: $(f \circ g)(x) = (g \circ f)(x) = x$.

inverse variation *See* indirect variation.

irrational number A number that cannot be expressed as a fraction, whose decimal form neither repeats nor terminates.

leading coefficient The coefficient of the first term of a polynomial once it's written in standard form; its term contains the variable raised to the highest power.

least common denominator The smallest possible common denominator for a group of fractions.

like radicals Radical expressions that contain matching radicands and indices.

like terms Terms containing variables that match exactly.

linear equation An equation of the form $ax + by = c$; its graph is a line in the coordinate plane.

matrix A group of objects called elements or entries (usually numbers) arranged in orderly rows and columns and surrounded with brackets.

mixed number A way to express an improper fraction that has an integer and fraction part written together like this: $5\frac{1}{2}$.

natural number A number in the set 1, 2, 3, 4, 5, ….

negative Describes a number less than 0.

number line Graphing system with only one axis, used to visualize inequalities containing only one unique variable.

numerator The top number in a fraction.

odd A number that is not divisible by 2.

one-to-one Describes a function whose inputs each have a *unique* output (no inputs share the same output value).

opposite Equals the given number multiplied by –1.

order Describes the dimensions of a matrix; a matrix with m rows and n columns has order $m \times n$.

ordinate Fancy-pants word for the y part of a coordinate pair.

parallel Describes nonintersecting lines; technically, they also have to be in the same plane, but all lines you draw in elementary algebra can be assumed coplanar. Parallel lines have the same slope.

perfect cube Generated by some quantity multiplied by itself twice; b^3 is a perfect cube since $b \cdot b \cdot b = b^3$.

perfect square Generated by some quantity times itself; a^2 is a perfect square since $a \cdot a = a^2$.

perimeter Distance around a two-dimensional object (the sum of the lengths of its sides).

perpendicular Describes lines which intersect at right, or 90-degree, angles; technically, they also have to be in the same plane, but all lines you draw in elementary algebra can be assumed coplanar. Parallel lines have slopes which are opposite reciprocals of one another.

piecewise-defined function Made up of two or more functions that are restricted according to input; each individual function that makes up the piecewise-defined function is valid for only certain x input values.

point-slope formula A line with slope m which passes through point (x_1,y_1) has equation $y - y_1 = m(x - x_1)$.

polynomial The sum of distinct terms, each of which consists of a number, one or more variables raised to an exponent, or both.

positive Describes a number that's greater than 0.

power *See* exponent.

prime A number or polynomial divisible only by itself and 1.

principal The amount of money you initially deposit in an interest problem.

product The result of a multiplication problem.

property A mathematical fact that is so obvious, it is accepted without proof.

proportion Equation that sets two fractions equal, such as $\frac{a}{b} = \frac{c}{d}$.

quotient The result of a division problem.

radical Symbol $\left(\sqrt{\ }\right)$ that, by its presence, defines a radical expression.

radicand In the radical expression $\sqrt[a]{b}$, b is the radicand.

range The set of possible outputs for a function.

rational number Describes something that can be written as a fraction (or a terminating or repeating decimal, if it's a number).

rationalizing the denominator The process of removing all radical quantities from the denominator of a fraction.

Rational Root Test Lists all of the possible fractions of the form $\pm\frac{C}{L}$, where C is a factor of the constant and L is a factor of the leading coefficient; that list represents all possible rational roots of the given equation.

real number Any number (whether rational or irrational, positive or negative) that can be expressed as a single decimal.

reciprocal Defined as one divided by the given number—the product of a number with its reciprocal always equals 1; the reciprocal of a fraction equals the fraction flipped upside down.

relation A rule that pairs inputs with outputs.

root (of an equation) *See* solution.

scalar A real number; usually used when discussing matrices to highlight the fact that the number is not a part of the matrix.

slope Number describing how "slanty" a line is; it's equal to the line's vertical change divided by its horizontal change.

slope-intercept form Describes a linear equation solved for y: $y = mx + b$, where m is the slope and $(0,b)$ is the y-intercept of the line.

solution Values that, if substituted for the variable(s) of an equation, make that equation true.

square matrix A matrix with an equal number of rows and columns; in other words, the matrix has order $n \times n$.

square root A radical whose index is 2.

standard form (of a line) Requires that a linear equation have form $ax + by = c$, where a, b, and c are integers and a is nonnegative (either positive or 0).

substitution Replacing a variable or expression with an equivalent value or expression.

surface area The amount of "skin" needed to cover a three-dimensional object, neglecting its thickness.

symmetric property If the sides of an equation are reversed, the value of the equation is unchanged; in other words, if $a = b$ then $b = a$.

synthetic division Technique for calculating polynomial quotients that's applicable only when the divisor is of the form $x - c$, where c is a real number.

system of equations A group of equations; you are usually asked to find the coordinate pair or pairs which represent the solution or solutions common to all the equations in the system.

terminating decimal A decimal that is not infinitely long.

terms The clumps of numbers and/or variables that make up a polynomial.

test point A coordinate used to determine which of the regions of a linear inequality or system of linear inequalities is the solution.

undefined Describes a number whose denominator is 0 but whose numerator is not.

variable Letter used to represent a number.

vertex (of an absolute value graph) Sharp point at which the graph changes direction.

vertical line test Tests the graph of a relation to see whether or not it's a function.

volume The amount of three-dimensional space inside an object.

whole number A number in the set 0, 1, 2, 3, 4, 5, ...

***x*-axis** Horizontal line on the coordinate plane with equation $y = 0$.

***y*-axis** Vertical line on the coordinate plane with equation $x = 0$.

zero product property If the product of two or more quantities equals 0, then one of those quantities must have a value of 0.

zeros (of a function) The x-values at which the corresponding $f(x)$ value equals 0; graphically speaking, the zeros are the x-intercepts of the function.

Index

Numbers

A

B

C

D

E

F

G–H

I–J–K

L

M

N

O

P

Q

R

W-X-Y-Z

Check Out These Best-Sellers

1-59257-115-8 • $16.95

1-59257-120-4 • $18.95

0-02-864456-5 • $9.95

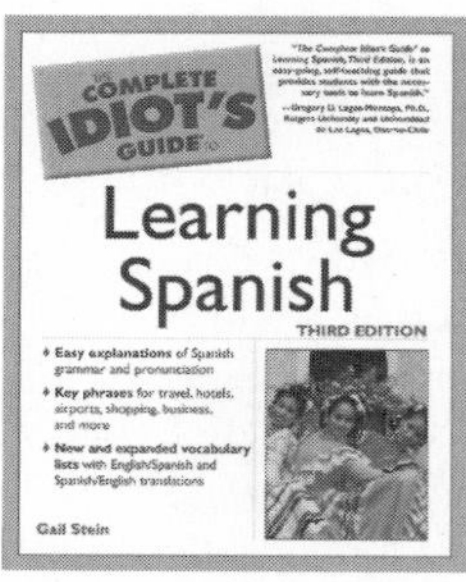

0-02-864451-4 • $18.95

0-02-864374-7 • $19.95

1-59257-413-0 • $16.95

1-59257-439-4 • $18.95

0-02-864463-8 • $14.95

1-59257-389-4 • $18.95

0-02-864365-8 • $18.95

1-59257-437-8 • $19.95

0-02-864440-9 • $14.95

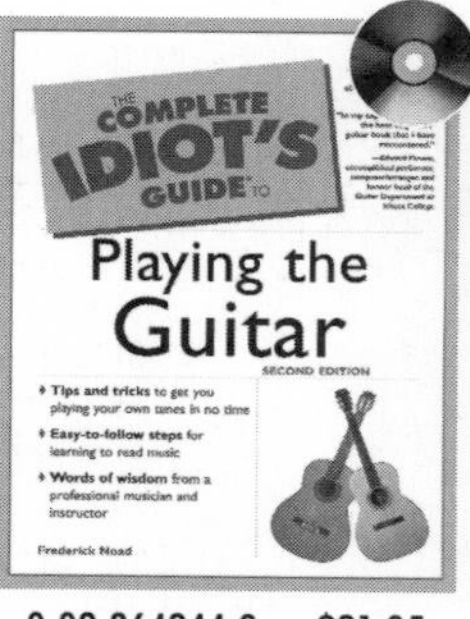

0-02-864244-9 • $21.95

1-59257-335-5 • $19.95

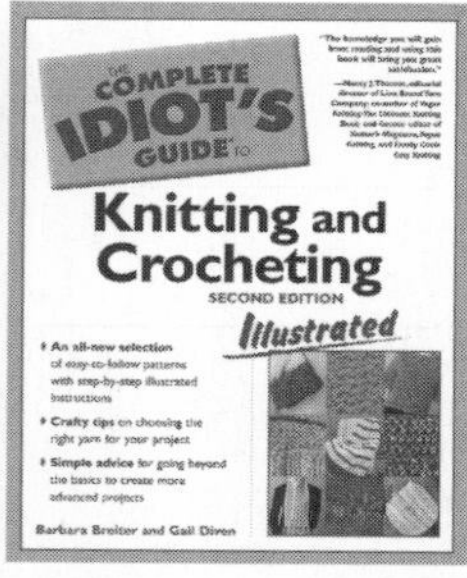

1-59257-089-5 • $16.95

More than *450 titles* available at booksellers and online retailers everywhere

www.idiotsguides.com